Liu Yang (Ed.)
Composites Recycling

Also of interest

Self-Reinforced Polymer Composites.
The Science, Engineering and Technology
Padmanabhan Krishnan and Sharan Chandran M, 2022
ISBN 978-3-11-064729-7, e-ISBN (PDF) 978-3-11-064733-4

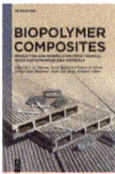

Biopolymer Composites.
Production and Modification from Tropical Wood and Non-wood Raw Materials
Edited by: Salit Mohd Sapuan, Syeed Saiful Azry Osman Al Edrus,
Ahmad Adlie Shamsuri, Aizat Abd Ghani and Khalina Abdan, 2023
ISBN 978-3-11-076919-7, e-ISBN (PDF) 978-3-11-076922-7

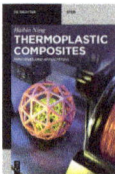

Thermoplastic Composites.
Principles and Applications
Haibin Ning, 2021
ISBN 978-1-50-151903-1, e-ISBN (PDF) 978-1-50-151905-5

Biopolymers and Composites.
Processing and Characterization
Edited by: Samy A. Madbouly and Chaoqun Zhang, 2021
ISBN 978-1-50-152193-5, e-ISBN (PDF) 978-1-50-152194-2

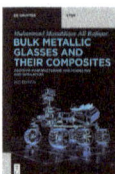

Bulk Metallic Glasses and Their Composites.
Additive Manufacturing and Modeling and Simulation
Muhammad Musaddique Ali Rafique, 2021
ISBN 978-3-11-074721-8, e-ISBN (PDF) 978-3-11-074723-2

Composites Recycling

Edited by
Liu Yang

DE GRUYTER

Editor
Dr. Liu Yang
Department of Mechanical and Aerospace Engineering
University of Strathclyde
75 Montrose Street
G1 1XJ Glasgow
United Kingdom
l.yang@strath.ac.uk

ISBN 978-3-11-075441-4
e-ISBN (PDF) 978-3-11-075443-8
e-ISBN (EPUB) 978-3-11-075453-7

Library of Congress Control Number: 2025938930

Bibliographic information published by the Deutsche Nationalbibliothek
The Deutsche Nationalbibliothek lists this publication in the Deutsche Nationalbibliografie;
detailed bibliographic data are available on the internet at http://dnb.dnb.de.

Preface

In recent decades, the increasing use of composite materials – particularly fibre-reinforced polymers – has transformed industries ranging from aerospace and automotive to wind energy, construction, and consumer goods. These materials have brought significant performance advantages due to their strength-to-weight ratio, corrosion resistance, and design flexibility. Yet with their widespread use comes a pressing challenge: how to sustainably manage their end-of-life and manufacturing waste.

Traditional disposal methods, such as landfilling and incineration, are environmentally unsustainable and waste valuable resources. These methods are increasingly becoming unacceptable in light of environmental regulations and circular economy goals. Without effective recycling solutions, composites industry is facing serious environmental, economic, and regulatory challenges.

Composites Recycling provides a timely volume delving into the complexities of recycling fibre-reinforced polymers, offering a comprehensive overview of current state of the field and a roadmap towards a circular materials economy. With contributions from leading experts in academia and RTOs, this book provides in-depth perspectives on mechanical, thermal, and chemical recycling methods, as well as the environmental implications of composites recycling.

Whether you are a researcher, engineer, policymaker, or student, *Composites Recycling* offers essential guidance for navigating this fast-evolving field.

Liu Yang
University of Strathclyde
November 2025

https://doi.org/10.1515/9783110754438-202

Contents

List of contributors

Prof. Albert ten Busschen
Windesheim University of Applied Sciences
T-Building, Campus 2
8017 CA Zwolle, The Netherlands
a.ten.busschen@windesheim.nl
Chapter 1

Dr. Jelle Joustra
Industrial Design Engineering
TU Delft
Landbergstraat 15
2628 CE Delft, The Netherland
Chapter 1

Alma van Oudheusden
Industrial Design Engineering
TU Delft
Landbergstraat 15
2628 CE Delft, The Netherland
Chapter 1

Dr. Matthew Keith
Chemical Engineering
University of Birmingham
Edgbaston
Birmingham
B15 2TT, UK
m.j.keith@bham.ac.uk
Chapter 2

Prof. Gary Leeke
Chemical Engineering
University of Birmingham
Edgbaston
Birmingham
B15 2TT, UK
G.A.LEEKE@bham.ac.uk
Chapter 2

Dr. Kyle Pender
National Composites Centre
Bristol and Bath Science Park
Feynman Way Central, Emersons Green
Bristol, BS16 7FS, UK
Kyle.Pender@nccuk.com
Chapter 3

Dr. Patric Sullivan
National Composites Centre
Bristol and Bath Science Park
Feynman Way Central, Emersons Green
Bristol, BS16 7FS, UK
Patrick.Sullivan@nccuk.com
Chapter 3

Dr. Fanran Meng
School of Chemical, Materials and Biological
Engineering
Sir Robert Hadfield Building
Mappin Street
Sheffield, S1 3JD, UK
fm392@cam.ac.uk
Chapter 4

https://doi.org/10.1515/9783110754438-204

Albert ten Busschen, Jelle Joustra, Alma van Oudheusden

1 Mechanical reuse and recycling of composites

1.1 Introduction

Composite materials in the context of a circular economy are a challenging subject. On the one hand, these materials allow for efficient materials use and long product lifetimes, on the other hand reprocessing at end-of-use (EoU) remains challenging. The majority of these materials are made with thermosetting polymers that cannot be melted and therefore the original components (fibres, resins) cannot be recovered without degradation. How to process composite products at the end of their use?

This chapter describes the mechanical methods for reuse and recycling of EoU composites and in particular structural reuse and mechanical recycling. The chapter starts with an overview of circular strategies for composite materials in Section 1.2. The subsequent sections describe pathways for mechanical reuse and recycling of composite materials. This starts with an explanation of the product build-up of the most important EoU composites in Section 1.3. Section 1.4 describes the first phase of mechanical processing and transportation of processed parts. Section 1.5 covers structural reuse and retrieving construction elements, followed by processing the material into oblong reinforcing elements in Section 1.6. Section 1.7 describes mechanical recycling of EoU composite products by grinding them into particles and fillers. The chapter concludes with a discussion (Section 1.8) and conclusions on mechanical reuse and recycling of composite materials.

1.2 Circular economy strategies for composites

The circular economy is an economic and industrial system, which aims to eliminate waste through restorative use of resources [1]. Resources are retained in the system rather than wasted by focusing on product, component, and material reuse [2, 3]. The circular economy strategies are prioritised according to their capacity to preserve embedded value and apply directly to the reusing of composite products or parts. Composite materials enable a long product lifetime because of the resistance to corrosion and fatigue [4, 5] and provide opportunities for lifetime extension through maintenance and repairs [5, 6]. Although the practice is not yet widespread, there are cases of successfully reusing products by refurbishing and remanufacturing of composite parts.

Routes for reprocessing materials into raw materials show some distinct aspects [4]. Composite materials differ from homogeneous bulk materials in the way they are processed and created, which directly affects the characteristics of the materials re-

https://doi.org/10.1515/9783110754438-001

covery processes. The majority of composite recycling processes tend to break down the composite into its constituent materials, thus losing the specific composite material properties [4, 7]. In addition to losing laminate architecture and fibre length, many recycling processes severely degrade materials, hampering the economic viability [7]. As a consequence, the majority of composite material goes to landfill or is incinerated, losing the material and its potential for reuse [8]. There is an urgent need for composite reprocessing solutions that can capture and retain material quality.

Structural reuse and recycling were identified as strategies to preserve material integrity. Structural reuse takes place through repurposing, resizing (segmenting), or reshaping the product. These actions discard the original product function, but maintain the unique structural properties, determined by the combination of material composition and structural design [9, 10]. Experts and the literature both note that this approach preserves material quality and value with a relatively small investment of energy and resources [5, 9, 11]. Keeping materials in use is an effective way to store resources for a longer time.

On a smaller scale, oblong elements like strips or flakes that are processed from EoU composite products can be used for making new composite products [12]. In that case, the elements are embedded in a virgin resin to form profiles or other structures with high percentages of retrieved content. The original composite structure contributes to the material characteristics of the new composite product. Although this method requires more processing steps, it offers more possibilities for reprocessing of large quantities [13].

Recycling of composite materials starts with mechanical downsizing into processable parts, followed by further processing. Thermoset composite reprocessing is usually based on polymer degradation, aiming to separate and extract fibres and resins. These processes can be generally classified as thermal, (electro-)chemical, or mechanical recycling. In mechanical recycling, addressed in this chapter, the composite is further processed into fragments or particle-based outputs.

Table 1.1 shows the overview of circular economy strategies for composite-containing products. These circular strategies are complementary to each other. For

Table 1.1: Circular economy strategies for composite materials.

Design aim	Preserving product integrity			Preserving material integrity	
Circular economy strategies	Long life	Lifetime extension	Product recovery	Structural reuse	Material recycling
Actions/processes	Physical-durability Long use Reuse	Repair Maintenance Adapt Upgrade	Refurbishment Remanufacture Parts-harvesting	Repurpose Resize Reshape	Mechanical Thermal Chemical Remould

example, a product can be reused and refurbished multiple times before reaching the end of use. Then, a composite part can find subsequent uses through repurposing, resizing, and recycling. The strategies presented in this chapter can be complimentary to each other as well as to those discussed elsewhere in this book.

1.3 Build-up of end-of-use composites

For effective mechanical reuse, a composite product waste stream has to meet a number of requirements. The material must be available in large volumes to set up an efficient recovery process. Collection and logistics have to be available and compatible with both material and processing facilities. Finally, the material must have residual mechanical properties that make it intrinsically viable for reuse. This pertains material characteristics at the end of use; thus load history and environmental degradation should be taken into account. In this section, the main sources of EoU composite material is first discussed, followed by the build-up of the three most relevant EoU product groups.

1.3.1 End-of-use volumes

Table 1.2 shows the main sources of EoU composite materials in the Netherlands and Europe. A significant volume of these sources is directly available for mechanical reuse. Wind turbine blades and polyester boat hulls represent the largest annual volumes, followed by roof panels and feedstock silos. The first two product types are widely recognised as significant EoU composite product streams. Various estimates have been made for the quantities of end-of-life wind turbine blades and boat hulls in Europe. It is estimated that the volume of blades reaching EoU – stage will increase from approximately 15,000 tonnes in 2019 to nearly 500,000 tonnes in 2050 [15]. For boats, less data is available, but it is estimated that between 120,000 and 145,000 tons of fibre-reinforced boats are in need of reprocessing per year in the EU [16, 17]. Roof panels and silos represent significant volumes in the Netherlands and given the widespread use of both, likely internationally as well. Printed circuit boards present a large volume but are, like other sources discussed in the next paragraph, unavailable for mechanical processing. Although the estimates vary in magnitude and timing depending on the assumptions made, it is clear that large quantities of composite materials need to be processed annually.

Various other composite applications result in large EoU product volumes but are unavailable for mechanical reuse. Printed circuit boards (PCBs), for example, are built on a substrate of glass fibre-reinforced composite. The recycling process of printed circuit boards is predominantly aimed at recovering precious metals rather than composite materials [19]. Also, approximately 145,000 ton of pipes and tanks are

Table 1.2: End-of-use composite in the Netherlands and Europe [14–18].

Composite product	EoU composite product (ton/year)	
	The Netherlands	EU
Rotor blades of windmills	1,300	15,000
Polyester boat hulls	1,400	120,000
(Corrugated) polyester roof panels	1,000	Unknown
Silos for feedstock of cattle	800	Unknown
Printed circuit boards	3,400	400,000
Total	**7,900**	**≥535,000**

produced annually in the EU, but these have not emerged yet as a significant EoU stream. Similarly, infrastructural composites (bridges and lock-gates) have been introduced in the market within the last two decades. Given the physical durability of composite materials in combination with the desired long lifespan of the design of such applications, these products are still in use and not yet available for waste supply. Other composite applications are too dispersed in the market, challenging efficient logistic collecting. For example, composites used in sport equipment (e.g. snow boards and bicycle frames), luxurious parts (e.g. special Jacuzzi's and yacht parts), building components (e.g. dormers and facades), and parts of cars, trains, and airplanes (e.g. skirts and stiffeners). Thus, a range of composite applications follow a different path of EoU treatment which makes them unavailable for mechanical reprocessing, even though these represent a significant waste stream. All these products and PCBs are therefore not considered in this chapter.

Most of the product groups presented in Table 1.2 are eligible for mechanical reuse. Wind turbine blades and polyester boat hulls are potentially suitable because of the high level of mechanical performance of the composite in these products. The same goes for silos. The properties of silos of over 40 years old are practically the same as when newly moulded. Polyester roof panels, however, are unsuitable because these panels are too thin-walled and degraded after many years of service. Eroded by sunlight and moisture they will therefore not be considered for mechanical reuse according to the method as described in this chapter. Wind turbine blades, polyester boat hulls, and silos as the main materials of interest for mechanical reuse and their build-up will be addressed in the following sections.

1.3.2 Build-up of wind turbine rotor blades

Wind turbines have grown considerably over the past decades, both in number and individual power. The hub height and rotor diameter increased steadily, enabled by developments in materials and engineering. In the early 1990s, a typical wind turbine would have a diameter of 20 m, and by 2010 this already grew to 100 m [20]. The largest blade produced measures 123 m in length [21]. Larger blades provide not only a larger swept area but also higher tip speeds. These increase the power generated by the wind turbine, but also the demand on the blade's structural properties.

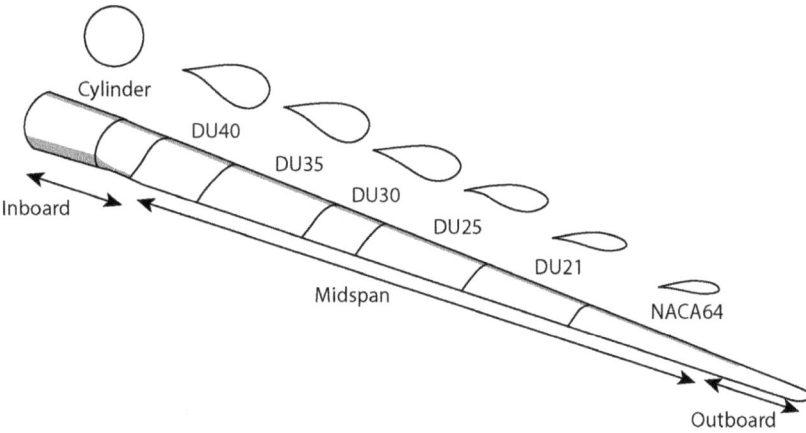

Figure 1.1: Schematic representation of NREL 5 MW blade, showing airfoils and main sections [10]. Reprint from Composites Part C: Open Access, 5, Jelle Joustra et al., Structural reuse of wind turbine blades through segmentation, 2021, with permission from Elsevier.

A wind turbine blade consists of three sections from root to tip: inboard, midspan, and outboard (Figure 1.1). The largest bending moment is exerted on the inboard section, where the blade is joined to the turbine axis. This is often a plain cylindrical shape made of a solid glass fibre-reinforced laminate. The midspan comprises approximately two-thirds of the blade length. A succession of airfoil profiles determines the blade's aerodynamic shape. The aerodynamic profile tapers towards the tip to meet aerodynamic and structural requirements. The outboard (tip) section has a relatively flat airfoil profile because it has to cope with high air speeds. In commercial blades, this section is often pre-bent to prevent collision with the tower when the blade deflects under load.

When zooming in on the blade cross-section, the integration of aerodynamic and structural design becomes more apparent. The shells on the leading edge (1) and trailing edge (2), as well as the shear webs (3) use a sandwich structure to provide high stiffness for minimum weight (Figure 1.2) [22, 23]. The sandwich combines triaxial glass fibre face laminates with a core material, such as a polymer foam or balsa wood. The thickness of the sandwich panels decreases along the blade length. Gradually, by lowering the number of plies in the layup (ply-drops), or stepwise by selecting thinner core materials. The spar caps (4) have a monolithic, unidirectional layup to provide longitudinal stiffness to the blade. Adhesive bonds (5) join the upper and

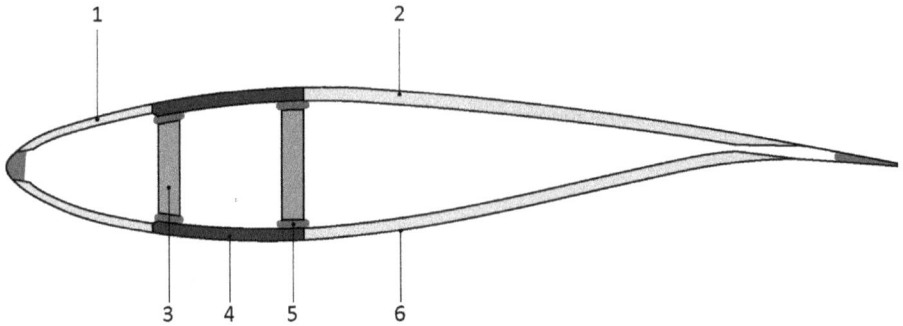

Figure 1.2: Sections in a blade cross-section: (1) leading edge shells, (2) trailing edge shells, (3) shear webs, (4) spar caps, (5) adhesive bonds, and (6) coating [24]. Reprint from Resources, Conservation and Recycling: 167, Jelle Joustra et al., Structural reuse of high end composite product: A design case study on wind turbine blades, 2021, with permission from Elsevier.

lower half of the blade, which are produced separately. A polyurethane surface coating (6) shields the materials from the environment and reduces wear.

Manufacturers use various material compositions, production technologies, and designs in their blades. While predominantly made of glass fibre-reinforced polymers, blades over 50 m in length also use carbon fibre [25] as unidirectional reinforcements in spar caps or hybrid weaves for the shells [23]. The variation is greater in terms of resins and clearly differs between manufacturers. Vestas uses epoxy, while LM Windpower mainly uses polyester [26, 27]. Covestro on the other hand develops and builds rotor blades based on polyurethane resins [28]. In conclusion, there is some degree of variation in compositions which may be relevant for reprocessing.

Technological developments in blade manufacturing are focused on increasing efficiency and lowering overall lifetime costs. While currently set at 25 years, some suggest extending wind turbine lifetime to 40 or even 50 years [29]. However, size and power increments described earlier also made repowering existing installations economically attractive [30, 31]. This means blades, or even complete wind turbines, may be replaced in favour of newer, more efficient models. Being replaced within their design lifetime, these blades are likely to be in sound physical condition when decommissioned.

1.3.3 Build-up of polyester boat hulls

Glass fibre-reinforced polyester (GFRP) has been the material of choice for many small watercrafts since the 1970s. The hull usually consists of a monolithic laminate reinforced with internal stiffeners [32]. The monolithic laminate of the hull is generally made of glass fibre-reinforced polyester. Various reinforcement types are used: random (laminated chopped strand mat or sprayed up mat), fabrics, and multi-axial reinforcement ("non-crimp fabrics"). The stiffeners are generally shaped on the wall-laminate using a core-material (most often foam based, but also other materials like wood, cardboard, or even old newspapers). Stiffeners can however also be pre-shaped and subsequently glued to the wall of the hull using a bonding paste.

Also, non-composite materials are incorporated into the boat hull and unfortunately in such a manner that they cannot be separated easily from the GFRP. These can be found below.

Coating outside: Onto the polyester "paint" on the outside of the boat (a so-called "gel-coat") often 2 K-PU-coating (DD-coating) are applied. In addition, an additional finishing layer of anti-fouling is often applied to prevent sea life (algae and molluscs) attaching to the hull.

In the wall of the hull: Inserts for connection purposes (metal, wood) and window-openings with rubber tightening can be found in the wall of the hull. (It is assumed that windowpanes have been removed.)

Inside surface: Boat interiors often use a polyester top-coating. Sometimes severe contaminations (grease, oil, etc.) are present on the inner surface that cannot easily be removed.

The polyester boat hulls have lengths that generally range from 4 to 9 m. The weight of GFRP of these hulls generally ranges between 400 and 3,000 kg per hull. As may be clear from the descriptions above, polyester boat hulls have very diverse build up and many other materials are connected to the GFRP of the hull. Sometimes also the deck is made of GFRP. Figure 1.3 gives an illustration of an abandoned polyester boat with an integrated GFRP deck.

Figure 1.3: Abandoned polyester yacht (obtained by ten Busschen).

1.3.4 Build-up of silos for cattle feed

Silos for cattle feed are present in the Netherlands in abundance (Figure 1.4). They all have a similar build-up but can be very different in size. The main body is a cylinder with a spherical cap on top and a conical bottom part for the outlet of feed. The silo diameters generally range from 2 to 4 m and a height ranging from 2 to 10 m. Volumetric capacities range from 4 to 120 m^3. The amount of GFRP in a silo (excluding steel pipes and steel support structure) ranges from 120 to 2,500 kg.

These silos are all built from GFRP. The cylindrical body is produced by quasi-tangential filament winding in combination with spray-up to achieve a wall-thickness ranging from 4 up to 10 mm. The spherical cap and conical bottom part are made by spray-up or hand lay-up. Tubes for filling and emptying may be of GFRP or of galvanised steel. The supporting frame is made of galvanised steel. The silo parts, support structure, and appendages are integrated by hand lay-up. Besides a relatively thin, UV-eroded outer surface these composites contain potential for structural re-use even after decades of outdoor exposure. Apparently, there is no structural degradation in the laminate. This is generally experienced by inspectors of old silos or producers of silos that get back old silos for refurbishment (e.g. Polem in the Netherlands).

Figure 1.4: Typical GFRP silos for cattle feed at a farm (with kind permission: taken from Polem, Industrieweg 7, the Netherlands).

1.4 First-phase mechanical processing and logistics

1.4.1 Logistics

For the EoU composite products considered for mechanical reuse, the first step is to transport the product from the location of use to a location where they are further processed. In the case of rotor blades there is a significant amount of processing necessary at the location of use before they can be transported to the location for further processing. The reason for this is that EoU rotor blades are very long (and will tend to be even longer in future) so transportation is very expensive. For polyester boats and silos for cattle feed this is generally not necessary as these can be transported on a truck without the need for processing first into smaller parts at location (Figure 1.5).

(a)
(b)

Figure 1.5: Transportation by truck of a silo (a) (with kind permission: taken from Polem, Industrieweg 7, the Netherlands) and boat hulls (b) (obtained by ten Busschen).

EoU boats, silos, and segments of wind turbine blades can be transported to the location for further processing. The costs are determined by the number of trucks and distance of travel. Generally, a truck will cost a certain rate in Euros per hour (including driver and excluding VAT) and can transport a certain amount of EoU material in kg. With these parameters, the costs for transportation of the EoU material can be calculated.

The transportation time will strongly depend on the distance but also on the time for waiting, loading, and off-loading. The amount of EoU material that can be transported by a truck can be increased by diminishing the volume. Especially boat hulls and silos have a large volume as compared to the weight. They are relatively thin-walled and can easily be broken into smaller parts with standard waste treating equipment on a truck.

Wind turbine blades however have a relatively high density and need specialised processing equipment to be broken into smaller parts. As already indicated, transport of long blades is expensive over land. Such costs may be justifiable at the installation phase, but not necessary in the EoU stage. Preferably, rotor blade parts should have a length of maximum 6 m. In general, there are four methods of cutting a rotor blade on site for making smaller segments: crushing, sawing, grinding, and water jet cutting.

1.4.2 Concrete crusher

A concrete crusher is a large, hydraulically driven tool that is generally mounted on a truck with a crane (Figure 1.6). Demolition companies use it for crushing concrete structures. They are also effective in roughly cutting a rotor blade in parts. However, a lot of dust and particles occur in the process, and these can be distributed in the environment by the wind. The use of a concrete crusher also gives severe material damage (cracks, delamination) in the material surrounding the cut.

Figure 1.6: Concrete crusher mounted on crane vehicle (with kind permission: taken from Boverhoff, Molenweg 1, the Netherlands).

There are certain measures that can be taken to prevent the distribution of dust by surrounding the workspace with containers, thus forming a wall, and by blowing a water mist on the area where the rotor blade is crushed to prevent dust pollution.

1.4.3 Diamond saw

Like a concrete crusher, a diamond saw is also mounted on a truck with a crane. A rotor blade can thus be cut in a controlled manner leaving the material of the rotor blade undamaged (see Figure 1.7). During cutting some dust is however generated. Often, water is added to the process for cooling. If this is the case, the dust is immediately entrapped in the processing water. In general, it is not possible to cut a rotor blade all the way through from one position, and additional cranes are necessary to turn the blade or hold it in the right position for cutting. The cutting width depends on the thickness of the diamond-tipped blade and is generally between 4 and 8 mm.

Figure 1.7: Cutting of a rotor blade with diamond saw (obtained by ten Busschen).

1.4.4 Grinding belt

A grinding belt is a dedicated machine where a long belt with grinding capacity can saw through a rotor blade from one position (Figure 1.8). Similar to a diamond saw, the material of the rotor blade remains undamaged and grinding dust will only appear at the location where the blade is being cut. Because the grinding belt is water-cooled, the dust will be directly entrapped in the process water. Other machinery is needed to bring the rotor blade in position allowing the belt to cut through the complete cross-section. The width of the cut depends on the cutting width of the grinding belt and is generally around 10 mm.

When a rotor blade is positioned within a small distance from the ground, a grinding belt can cut a rotor blade into smaller segments without any further handling during the operation. For a medium size rotor blade (e.g. 40 m length) a cross-cut is made in 3–8 min, depending on the location in the length direction of the rotor blade.

Figure 1.8: Grinding belt (with kind permission: taken from Odolphi Technical Support, Marnchiem 14, the Netherlands).

1.4.5 Waterjet cutting

Water cutting, or "cold cutting" is done by projecting a high-pressure waterjet on the object to be cut. The pressure is very high, generally 3,000 bars and the water contains sand as an abrasive. Using water cutting practically all materials can be cut. To cut a rotor blade a mobile set-up is necessary. In view of safety, a guiding frame is first installed onto the rotor blade that can guide the cutting head in a controlled manner. As the width of the cut is very small with water cutting, generally 0.2–0.4 mm, very little dust is generated. Moreover, by nature of the process, this dust is directly entrapped by the processing water. Water jet cutting and a grinding belt both cut a rotor blade into neat sections, as shown in Figure 1.9.

In Table 1.3 an overview is given of some characteristics of the methods for on-site cutting of rotor blades. In all cases dust is generated, either entrapped in processing water or not. It is important that measures are taken to catch the generated materials and remove them to prevent environmental pollution.

Figure 1.9: Blade segments after grinding with belt (with kind permission: taken from Odolphi Technical Support, Marnchiem 14, the Netherlands).

Table 1.3: Characteristics of methods for on-site cutting of rotor blades.

Method	Dust generation	Particle generation	Product damage
Concrete cutter	Severe (*)	Severe (*)	Severe
Diamond saw	Moderate, controlled	Practically none	Not
Grinding belt	Moderate, controlled	Practically none	Not
Water cutting	Very little	Practically none	Not

*As mentioned earlier, the generation of dust and particles in case of the use of a concrete cutter can be controlled by measures at the working site. By using a water mist, a container wall around the workspace and cleaning up the water with the dust and the particles from a tight workspace floor the pollution of the environment can be prevented. This is brought in practice by e.g. Lubbers Logistics Group in the Netherlands.

1.5 Structural reuse of parts

Structural reuse is particularly interesting for composite materials as it preserves the original material composition and characteristics. Structural reuse can take place through repurposing, segmentation/resizing, or reshaping the product. These actions discard the original product function, but maintain the unique structural properties, determined by the combination of material composition and structural design. Structural reuse differs from product recovery in the sense that the parts are reused in other functions then initially designed for. Furthermore, structural reuse preserves material integrity, but differs from recycling processes in how it retains the original material structure. Structural reuse is therefore positioned as circular strategy between product recovery and materials recycling (Table 1.1).

1.5.1 Repurposing

Repurposing revolves around reusing the product as-is, or large parts of it, in another function. The composition of the material and product, as well as the overall geometry, is retained. Relatively few interventions are needed to make the product suitable for reuse. However, this does mean that the secondary application depends on the geometry and layup designed for the first application. On the one hand, this presents a challenge to match those applications as closely as possible in order to make the best use of the material. On the other hand, over-dimensioning secondary applications (i.e. using excess material to warrant a high safety margin) could even be considered desirable, as this increases the volume of materials being reused and "stored" in active stock.

(a)　　　　　　　　　　　　　　　(b)

Figure 1.10: Bridge of blades, using two decommissioned wind turbine blades as load carrying beams [33] (With kind permission: created by Stijn Speksnijder, Trompstraat 334 G, the Netherlands).

Recent publications show a range of wind turbine blade repurpose concepts, where infrastructural applications play a major role. The most notable examples are bridges [33, 34], power transmission line poles [35], playgrounds, and sound barriers [36]. Some of these have been realised as demonstrator objects.

Bridges present a compelling reuse case for wind turbine blades. The Bridge of Blades (Figure 1.10) uses 75–85% of the blade length as load carrying beams [37].

Other studies report reuse percentages starting at 40%, depending on required span and load case [35]. To support engineering of blade bridges, Swedish research institute RISE developed a calculation tool and database to match bridge design criteria such as span and required stiffness with blade availability [38]. It is expected that the residual mechanical properties suffice for the loading scenario in a slow traffic bridge and will, as in a conventional steel bridge, have a lifespan of 60 years [34]. Such pedestrian and cyclist bridges are currently being constructed in Ireland and Poland [39].

To implement repurposing on a large scale, there are still a number of challenges related to the reverse supply chain and design. The reverse supply chain is currently still underdeveloped, which means there are still major challenges regarding the variation in materials supply (models as well as planning), the logistics associated with transporting large-scale parts, and reprocessing technology. In terms of design, there are challenges to properly deploy recovered materials and to anticipate repurposing as part of the product lifecycle in the design of the next generation of composite products. Although it is attractive to reuse large parts, finding suitable applications remains challenging.

1.5.2 Resizing and segmentation

To match the supply of EoU materials, recovery should be done on a systematic basis, preferably in larger series, or mass-production. Cutting larger composite parts, such as silos and blades, into practically usable construction elements is expected to expand the range of potential reuse applications. Construction elements like panels and beams are found across many applications in building and construction, infrastructure, and furniture industry. Initial experimentation showed that several high valued objects can be made, as furniture for example [24]. This requires a two-step approach (Figure 1.11). First, the structure is segmented into reusable construction elements like panels and beams. Second, the obtained elements are used in a next product lifecycle.

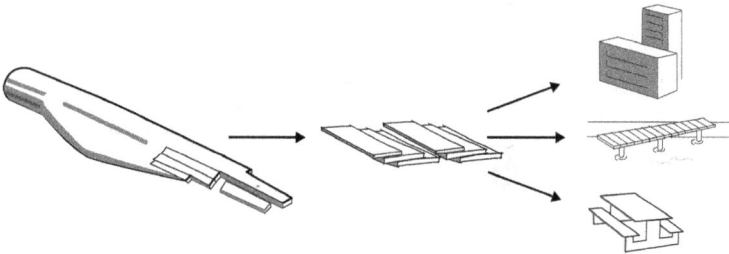

Figure 1.11: Structural reuse of a wind turbine blade: segmentation into construction elements and reuse in diverse applications [24].

The properties of the retrieved elements depend on the way they are cut from the original product. For the case of wind turbine blades, a number of cutting patterns were developed. Joustra [10] determined a cutting pattern based on timber standards and estimated that 55% of the blade midsection could be directly reused as panel or beam segment. The root and tip sections were not considered because of the thick laminate structure and strongly (double) curved surfaces. Another approach would be to section the blade into narrow strips or slats. Cutting into narrower strips has the advantage that curvature plays a relatively minor role and will thus increase the re-covery percentage. However, it also results in narrow boards with a narrower scope of reuse. To better suit the construction industry, Pronk [40] therefore developed a cutting pattern based on standard timber trade dimensions. This segmentation proce-dure achieves recovery up to 60 wt.% (Figure 1.12). Depending on the chosen cutting pattern, the majority of a blade could thus be cut into panel or beam element.

(a)

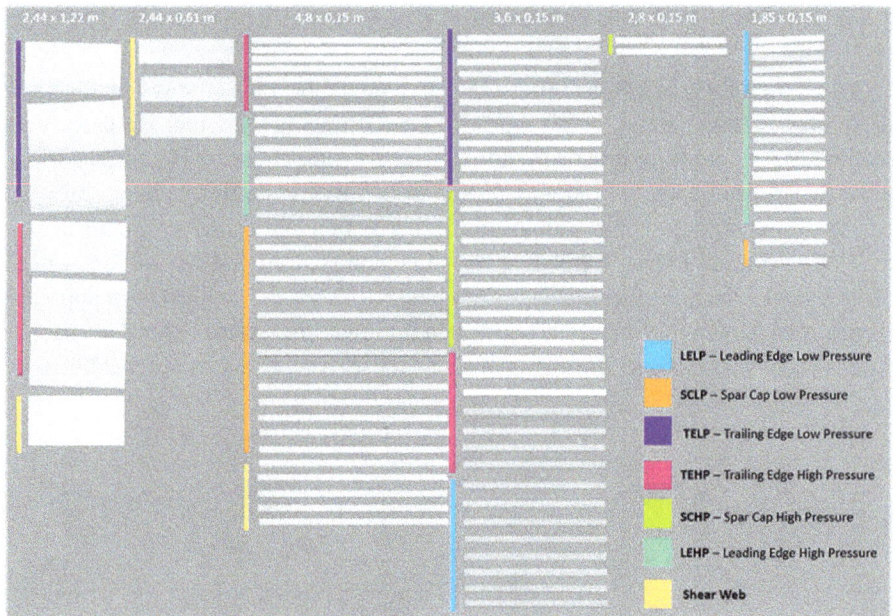

(b)

Figure 1.12: (a) Cutting pattern and (b) obtained standard-sized panels and beams from a wind turbine blade [40] (with kind permission: created by Simon Pronk, Lisstraat 15A, the Netherlands).

Several application areas were identified for recovered construction elements, including construction and transport applications. In construction, the recovered elements could be used as structural parts in flooring, walls, or roofing. Sourcing from secondary feedstock and using elements with high specific mechanical properties, such as lightweight panels, provides interesting advantages in construction and transport. Structural reuse would make such high-end composite materials available at a favourable price. In transport, smaller planks and recovered elements could be used for lightweight, yet strong packaging, such as pallets. Both product groups can be produced centrally with appropriate processing equipment and dust extraction.

Figure 1.13: Design demonstrator: furniture made of wind turbine blade panels [24]. Reprint from Resources, Conservation and Recycling: 167, Jelle Joustra et al, Structuralreuse of high end composite product: A design case study on wind turbine blades, 2021, with permission from Elsevier.

Design case studies and demonstrators showed the feasibility of the structural reuse for wind turbine blades. The blade's aerodynamic profile causes minor curvature, twist, and bend of harvested segments. However, at the scale of construction elements, these were found to be well within conventional construction tolerances [10]. To test this, retrieved segments were used to construct a simple furniture product (Figure 1.13), which was presented at a design exhibition. Curvature and origins of the panels went unnoticed to most visitors, until it was pointed out to them. In the discussions that followed, the material was appreciated as construction element [24]. The segmentation and reuse approach are expected to be applicable to other composite structures as well. Wind turbine blades represent a durable, structural composite, which integrates mechanical performance and aerodynamic form. In practical terms, it is a large product that has no other functional use when its operational life ends. Similar characteristics are found for composite products in other sectors, such as boat hulls and silos. As such, it is to be expected that structural parts could also be reused in these product categories.

1.5.3 Re-shaping

Structural reuse is not restricted to thermoset composites. Thermoplastics based composites offer additional opportunities because they can be remoulded. Thermoplastic composites are often produced as flat panels (blanks), which are subsequently thermoformed into a product shape. The forming process forces fibre, resin, and plies to move into the composite, affecting its material properties. At the end of use, a similar process could be used to adapt a given shape to meet the requirements of a secondary use case. It is expected that such subsequent forming actions widen the scope for reuse applications, as the material is no longer bound to its original shape, which would have been the case with thermoset resins.

The re-shaping process is particularly applicable to panel-type products [41, 42]. Kiss [41] demonstrated flattening of a stamp-formed thermoplastic composite object to a panel. Retaining fibre orientation proved to be a challenge as it is difficult to effectively reverse the original forming process. Cousins [42] demonstrated a re-shaping process on an experimental wind turbine blade spar cap. The spar cap was flattened through thermoforming. It was suggested that such flattened panels could be planed to a desired thickness for use in construction or recreational goods. The re-shaping process is still highly experimental and needs further research to arrive at a scalable recovery process.

1.5.4 Material performance

Construction materials obtained by segmenting composite structures into smaller elements have favourable characteristics. Joustra et al. [10] modelled the material characteristics of panel and beam elements retrieved from a wind turbine blade. Regarding flexural modulus, the elements retrieved from a blade shell are comparable to wooden panels (Figure 1.14). On flexural strength, the panels perform similar to aluminium or even steel. The density of these panels, however, is relatively low compared to metals, partially because of the use of low-density core materials in a sandwich layup. Altogether, this makes for lightweight, yet strong and stiff panels, which outperform conventional construction materials when loaded in bending [10].

Wind turbine blades are generally considered safe to use in construction [9]. The secondary load case is often well within the original design specifications. However, cutting the original structure into reusable segments exposes core materials and otherwise enclosed interior parts to environmental conditions. To prevent ageing from UV-radiation and moisture ingress, suitable precautions must be taken. Surface treatments will generally suffice to protect the materials and prevent degradation [43].

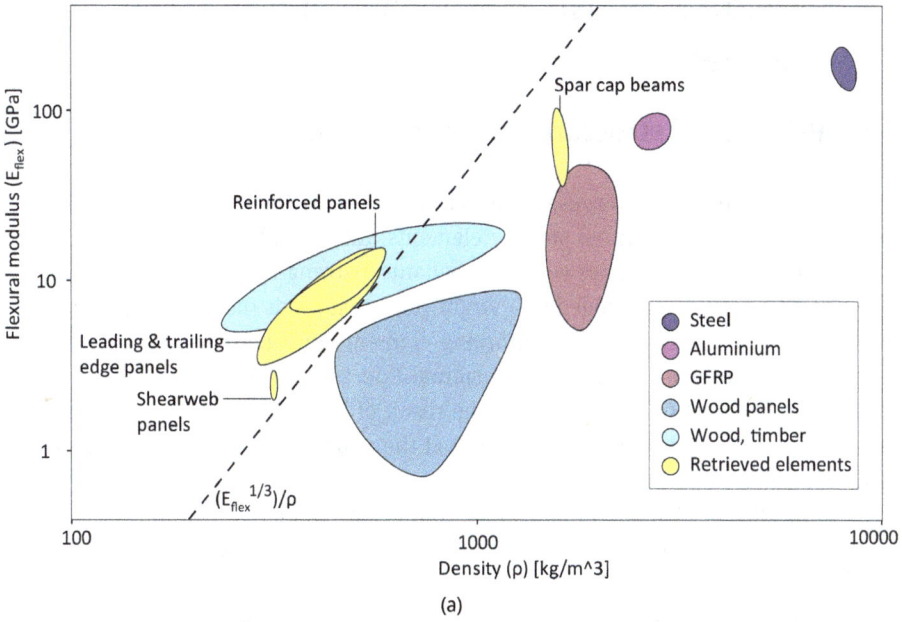

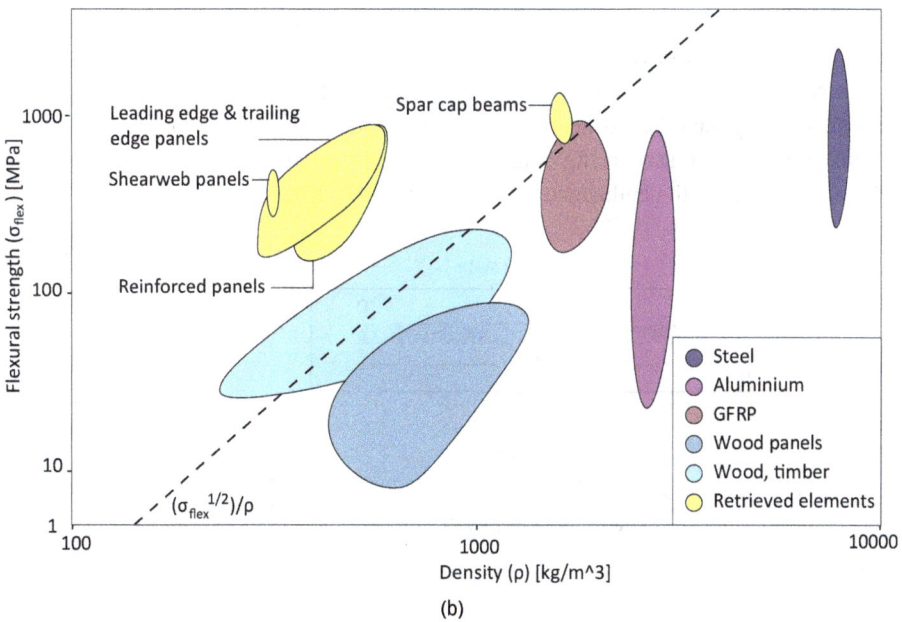

Figure 1.14: Material charts of density versus flexural modulus (a) and strength (b) of retrieved wind turbine blade materials [10]. Reprint from Composites Part C: Open Access, 5, Jelle Joustra et al, Structural reuse of wind turbine blades through segmentation, 2021, with permission from Elsevier.

1.6 Structural re-use of mechanical processed material

1.6.1 Principle of structural re-use with oblong elements

In Section 1.5, the structural reuse of parts is described. When EoU composite products are machined into substantial smaller elements, these small elements still have the intrinsic composite properties such as mechanical strength and resistance to corrosion [12]. They can be embedded in a virgin resin that binds the parts together. However, to use these elements as reinforcing elements in a new product and benefit from the mechanical strength of the original EoU composite product, the shape of these elements must be oblong. This is the result of the necessity of having sufficient length to load the ends of the element so that the middle of the element will have the same deformation (strain) as the surrounding material.

The mechanical loading of embedded reinforcing elements with finite length has already been described in 1952 with the shear-lag theory [44]. When an oblong, reinforcing element is embedded in a larger body that is strained, the part will build up the strain from the ends by shear stresses. When a part is considered, e.g. a strip with length L and thickness D that is embedded in a matrix material, the strain of the combination can be analysed. The shear-lag theory describes the occurrence of shear stresses, τ, as a result of the difference between strain deformation of the combination through an externally applied composite stress, σ_c, and the strain in the embedded element, resulting in tensile stresses in the embedded element, σ. The shear stresses enable the build-up of strain of the embedded element starting from the ends (Figure 1.15).

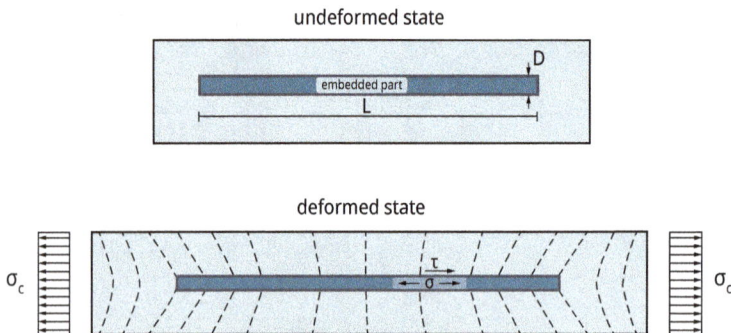

Figure 1.15: Loading of embedded element by shear deformation at the ends (obtained by ten Busschen).

From the figure it is clear that there is a transition region at the ends of the embedded element where the embedded element is not completely strained to the level of the composite strain. Over this region the tensile stress in the embedded element, σ, will increase from the externally applied composite stress, σ_c, to a maximum stress level in the middle of the embedded element. Figure 1.16 shows the distribution of shear stresses (τ) at the surface of the embedded element and normal stresses (σ) inside the embedded element.

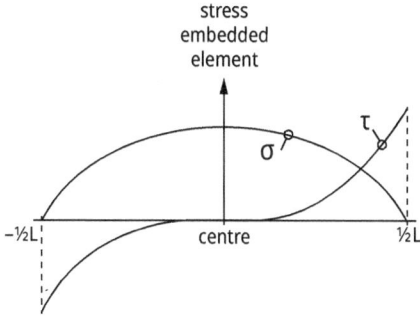

Figure 1.16: Distribution of shear stresses and normal stresses in the embedded element (obtained by ten Busschen).

1.6.2 Effect of L/D-ratio on composite properties

From literature it is extensively known that for composites reinforced with fibres of short length, the ratio between the (average) fibre length, L_f, and the fibre diameter, D_f, has a large influence on mechanical properties of the composite. Ning et al. [45] presented an elaborate literature review on this subject. The effectiveness of the reinforcement with fibres of short length can be studied for different mechanical performance parameters of the composite such as stiffness (E-modulus), static strength, impact strength, creep resistance, and fatigue strength. A specific mechanical performance parameter for a composite made with fibres of short length, P_s, is generally compared with the mechanical performance using endless (continuous) fibre reinforcement, P_c. An efficiency parameter, η, is introduced as the ratio between the mechanical performance parameter for a composite with short fibres and with continuous fibres:

$$\eta = P_s/P_c$$

The influence of the L_f/D_f-ratio on the efficiency of mechanical performance of the composite is very dependent on the mechanical performance parameter analysed. When plotting the efficiency parameter as function of the L_f/D_f-ratio on a logarithmic scale the following results are obtained for a variety of mechanical properties as schematically depicted in Figure 1.17.

For structural re-use of mechanically processed composite the effect of length to thickness on mechanical performance of a new product has been investigated. In the investigation a series of composite products were made that were built up from rectangular composite strips with a defined ratio between the strip length (L) and the thickness of the strip (D) [46]. New composite profiles were made with L/D-ratios of 8, 16, 40, and 200, respectively. The profiles were tested in three-point bending and

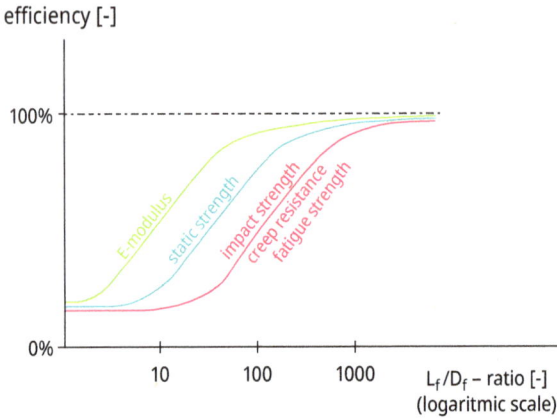

Figure 1.17: Typical efficiency curves for different mechanical performances (obtained by ten Busschen).

the effective stiffness (E-modulus) and effective bending strength were determined. Figure 1.18 shows the relation between the E-modulus and L/D-ratio. The mean value of the E-modulus ranges from 8.5 GPa for the lowest L/D-ratio to 9.5 GPa for the highest L/D-ratio. It is clear that considering the scatter in test results the dependence of the bending stiffness from the L/D-ratio is not very strong.

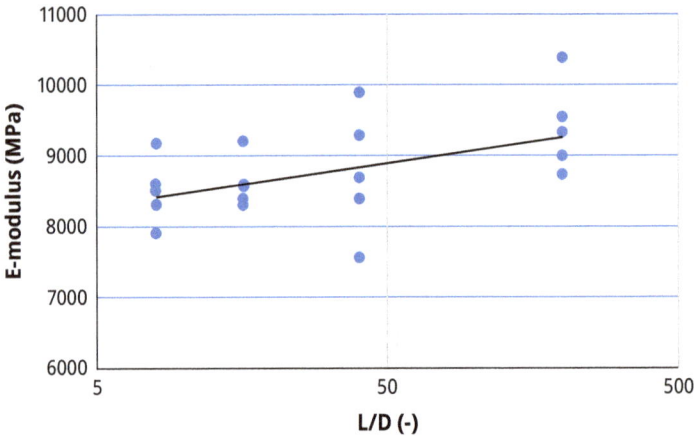

Figure 1.18: E-modulus of new product as function of L/D-ratio of strips [46] (obtained by ten Busschen).

Figure 1.19 shows the relation between the bending strength and L/D-ratio. The mean value of the strength ranges from 80 MPa for the lowest L/D-ratio to 200 MPa for the highest L/D-ratio. Clearly the effect of L/D-ratio on strength is strong. For a significant strength contribution of the reinforcing strips to the strength of the new product the L/D-ratio should preferably be 50 or higher.

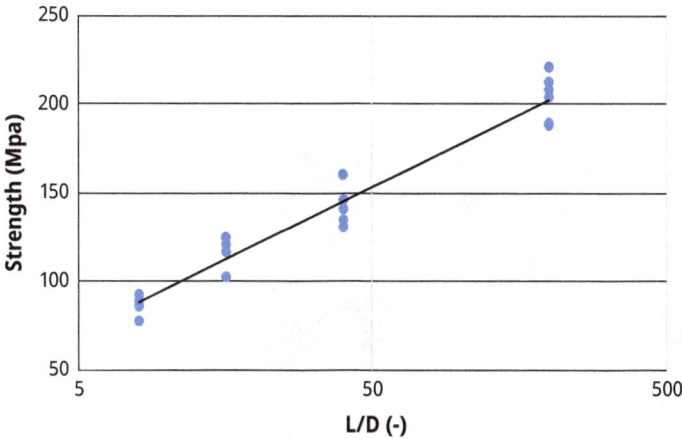

Figure 1.19: Strength of new product as function of L/D-ratio of strips (ten Busschen).

1.6.3 Processing into reinforcing elements

From composites that were exposed to outdoor conditions it is known that mechanical performance is preserved for a long time, generally much longer than the time that they are in use [47]. Thus, when EoU composites are processed into oblong reinforcing elements, there is strong potential to contribute to mechanical stiffness and strength in new products in the case of oblong reinforcing elements with a sufficient L/D-ratio. For significant strength contribution, the L/D-ratio should preferably be 50 or higher. Processing of EoU composite products into strips is in theory the best way to preserve and re-use mechanical performance of the composite. It was investigated how to process polyester boat hulls into strips and re-use them for new products.

Before a boat hull can be machined into strips, the hull must be machined into pieces that can be sawn. Figure 1.20 shows the breaking of a boat hull into such pieces. The necessary equipment for this process can be present on the truck used for transportation. With such equipment, polyester boat hulls can be broken into pieces with a surface-dimension in the order of 0.5–1 m². The complete operation is completed within 15 min for a medium-sized hull.

After the initial processing of polyester boat hulls into pieces typically in the order of 0.5–1 m², they can be sawn into strips using a diamond-tipped saw. Figure 1.21 shows plates obtained by tearing a boat hulland the resulting strips after sawing.

As can be seen in Figure 1.21, the resulting strips are curved. For small-scale production and testing, these strips can be processed into a new product but to date the processing of strips of EoU composites into new composite products was unfortunately not found to be suited for industrial, efficient production.

Figure 1.20: Photo of breaking a boat hull in smaller parts (obtained by ten Busschen).

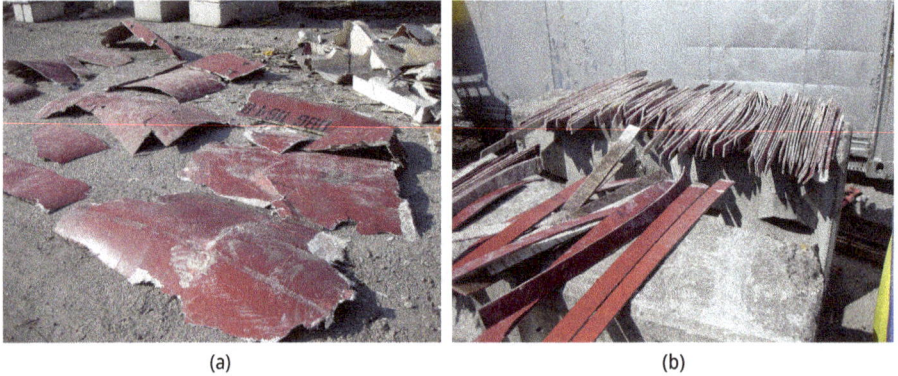

(a)   (b)

Figure 1.21: Plates from a boat hull (a) that are sawn into strips (b) (obtained by ten Busschen).

A better method for processing of EoU composites is shredding. A shredding process produces flakes with the required oblong shape. The advantage of flakes above strips is that they are smaller in dimensions and have a uniform distribution in size and shape. However, the requirement is still valid for these flakes to have a sufficiently high L/D-ratio to contribute as reinforcement.

Generally, shredding is carried out with a beater-mill (see Section 1.7 of this chapter) and the flakes can be sieved off when they are sufficiently small in dimensions. The drawback of shredding is the generation of dust. In case of a beater-mill, about

30 wt.% of the original EoU composite parts that are fed into the beater-mill is transformed into dust. This dust may be polymer-rich and can be used for pyrolysis into chemicals. Another outlet for the dust is to use it as filler in resins to make new composite products. The use of dust originating from EoU composite as filler is discussed further in Section 1.7.

The geometry of the flakes depends on the original EoU composite material. When this material consisted of random reinforcement (as is often the case for boat hulls), the flakes are plate-shaped, see Figure 1.22 left. When the original EoU composite had reinforcement with high orientation (as is generally the case for rotor blades), the flakes are spike-shaped, see Figure 1.22 right. Shredding silos will give a mixture of the two types of flakes because these products are composed of a combination of continuous filament-wound reinforcement and random reinforcement by spray-up.

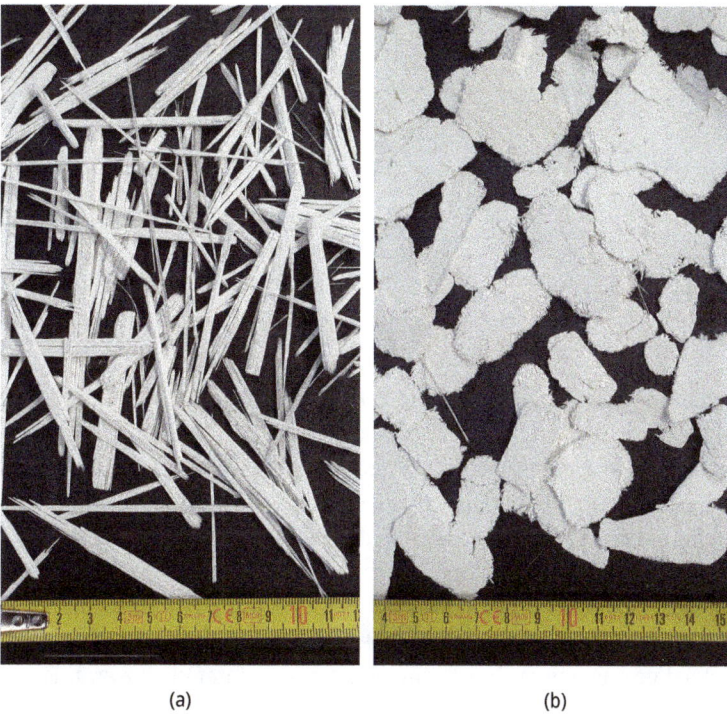

(a)   (b)

Figure 1.22: Flakes originating from boat hulls (a) and from rotor blades; (b) (obtained by ten Busschen).

1.6.4 Processing of flakes in new products

When the flakes are to be used as reinforcement in new products, they have to be bound together by a virgin resin. It was found that profiles or other products only consisting of flakes embedded in resin will show strength lower than expected. This is caused by the influence of the flakes at the surface of the product, which cause stress concentrations and crack initiation. To make profiles or beams based on flakes, the most optimal cross-section consists of a core of flakes embedded in virgin resin and an outer layer of virgin GRP. Using UD-reinforcement in this virgin GRP-layer, the mechanical properties can be tuned to the desired level of performance. Moreover, the use of UD-reinforcement can enhance creep resistance and fatigue strength significantly even when present in a relatively small amount. Figure 1.23 shows the principle of an optimum design of a cross-section.

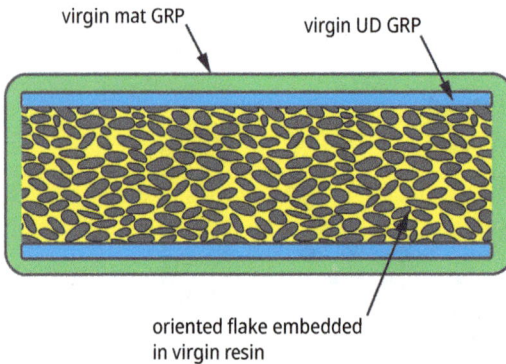

Figure 1.23: Principle of cross-section of optimum profile with EoU composite (obtained by ten Busschen).

A profile can be manufactured in a discontinuous manner. This can be done using an open mould in which the dry components are placed, followed by casting of a low-viscous resin. This is a good method for small-scale production and tests but not suitable for industrial production. Both labour costs and the costs for virgin resin become too high for a profitable business case because with this method over 50 wt.% of the total product is formed by virgin resin. Another discontinuous method is to use resin infusion methods such as vacuum infusion under foil, Vacuum Assisted Resin Transfer Molding (VARTM) or Resin Transfer Molding (RTM). Again, these methods are not suitable for industrial production. Although the virgin resin content is lower, it is still around 50 wt.% of the total product and too labour-intensive [12].

An industrial method to produce profiles or beams from flakes of EoU composite product is to use a so-called push-pultrusion process [46]. This is a continuous, pultrusion-like process pushing a core of oriented flakes between virgin reinforcement layers. This combination contains a surplus of virgin resin that gets pulled through a heated mould. The surplus resin is squeezed out and the package cures in the heated mould into a product. A schematic overview of the process is given in Figure 1.24.

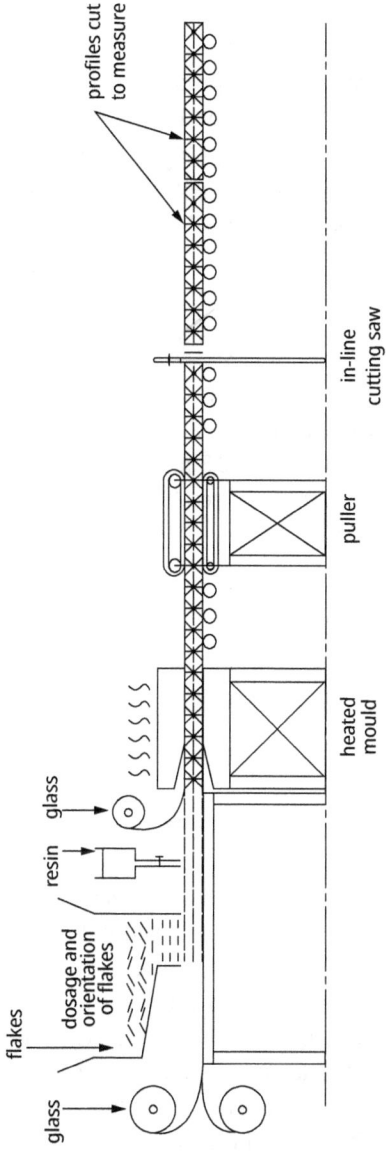

side view push-pultrusion re-used composite

glass

flakes

resin glass

dosage and orientation of flakes

profiles cut to measure

heated mould

puller

in-line cutting saw

Figure 1.24: Schematic overview of push-pultrusion process (obtained by ten Busschen).

1.6.5 Demonstrators

Structural re-use of EoU thermoset composites enables products that are suitable for infrastructural applications. Panels, profiles, or plates can be manufactured that are strong and resistant to moisture. That these new composite products are not light weight is, generally speaking, not a problem in these applications. Demonstrator infrastructural projects from re-used EoU composite products are presented in the next sections.

1.6.5.1 Retaining walls

EoU composites profiles have been manufactured into retaining walls close to a lock-gate in the Netherlands [46]. Profiles with a length of 3.5 m have been produced by using vacuum infusion. The profiles cross-section measures 40 × 250 mm with a detailing that profiles fit in each other (see Figure 1.25).

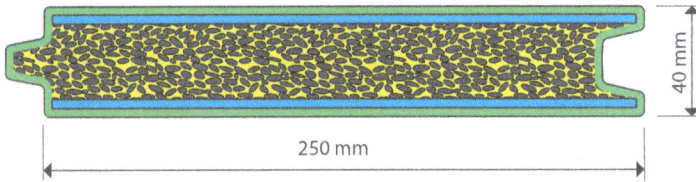

Figure 1.25: Cross section of retaining wall profile from EoU composite (obtained by ten Busschen).

The profile contributes to the wall with an effective width of 250 mm and has a thickness of 40 mm. The outside skin consists of a glass-reinforced polyester laminate made of virgin materials, containing a glass mat (in the picture-coloured green) and UD-reinforcement with 200 mm width on both sides (in the picture-coloured blue). Both reinforcements have a weight per unit area of 900 g/m². Inside the profile is filled with re-used EoU thermoset flakes embedded in a virgin polyester resin.

Prior to the vacuum infusion process, first the glass injection mat has been placed along with the first layer of UD-glass reinforcement in the open mould. On this, EoU flakes and strips have been deposited. A second layer of UD-reinforcement was placed on this and finally the glass injection mat was closed above it. After this, foil was place on this and made air-tight to the flange of the mould. Injection has been performed, using a polyester injection resin, with brown colour to give the profile a wood-like appearance. Further details can be found in the process description by ten Busschen [46]. Figure 1.26 shows the deposition of strips and flakes of EoU composite in the mould (left) and the product with infused with brown resin (right).

(a)   (b)

Figure 1.26: Production of profiles using EoU composite by vacuum infusion under foil (obtained by ten Busschen).

The profiles have been tested in bending to confirm that the profiles have an equal bending strength as for a profile made of tropical hard wood that is normally used in these applications (azobé). The retaining walls were installed by driving the profiles into the soil using a vibrating hammer (see Figure 1.27). During the vibrating process no damage was incurred. Profiles were withdrawn two years after installation and showed no degradation or strength loss after inspection and tests.

1.6.5.2 Guiding beams

Guiding structures are placed in canals near bridges or lock gates to guide ships. Traditionally, these guiding structures are built up from frame of steel on which guiding beams are mounted horizontally. Traditionally these beams are made of tropical hard wood with typical dimensions 200 × 200 × 4,000 mm. In the project, guiding beams of re-used EoU thermoset composite were commissioned. These were used to replace the lowest two rows of four guiding structures, located near a bridge in the Netherlands [46]. The bottom rows of guiding beams are around and under the water level and therefore this is where tropical hardwood beams suffer most from rot. Re-used EoU thermoset composite will be far more durable at this location. Figure 1.28 shows one of guiding structures. The upper four rows of beams are made from tropical hardwood, and the lowest two rows are made or re-used EoU composite.

(a)

(b)

Figure 1.27: Retaining wall installation by vibrating EoU composite profiles in the ground: (a) close up and (b) in distance (obtained by ten Busschen).

Figure 1.28: Guiding structure beams made from EoU composite (obtained by ten Busschen).

The guiding beams are designed to resist a possible ship collision. For this, a single beam has to have a static strength of 440 kN when loaded in bending when supported with a distance of 1,800 mm. The amount of UD-glass reinforcement in the beam was optimised to achieve this strength. A prototype beam was tested in three-point bending, supported with a distance of 1,800 mm. A maximum force of 515 kN was attained. Passing the strength test, guiding beams have been produced for the project with a total length of 112 m. This was done with RTM injection with an open steel mould with an upper composite mould part. Installation of the beams at the bridge in Delfzijl took place in October 2019.

1.6.5.3 Crane mats

Crane mats are used at building sites, creating a stable work area. Heavy cranes and other machinery can operate on this area. Crane mats generally are made of tropical hard wood beams, based on the requirements on strength, wear resistance and durability. Crane mats shown in Figure 1.29 have outside dimensions of 1,000 × 5,000 mm and are assembled from five beams of tropical hard wood that have dimensions 200 × 200 × 5,000 mm. The assembly is made using five steel bars, clamping the beams together.

Figure 1.29: Work place created with crane mats.

A crane mat was made of EoU thermoset composite beams. This was done in cooperation with Welex in the Netherlands. This company is a manufacturer of crane mats. For the production of beams made of EoU composite, RTM-infusion was used [46]. The process started with a glass injection mat (900 g/m^2) that was placed in the mould. This was followed by two layers of quadraxial glass reinforcement (1,200 g/m^2) and 10 layers of UD glass reinforcement (840 g/m^2). On this package flakes of EoU composite in combination with dry sand (1–2 mm grain) was deposited. The five holes with diameter of 30 mm for crane mat assembly were in-situ formed in the beams using inserts.

On the core another 10 layers of UD glass reinforcement was placed and the quadriaxial reinforcement layers and injection mat were folded over the top. Then the mould was closed for starting the resin injection process. With this RTM-infusion process, five beams have been produced and were assembled into a crane mat with traditional build up and dimensions. The crane mat was tested in various severe operating conditions. The EoU composite crane mat showed very good performance and good resistance to heavy vehicles and was easy to clean (Figure 1.30).

Figure 1.30: Crane mat of EoU composite tested with heavy vehicle (obtained by ten Busschen).

1.6.5.4 Bridge deck profiles

A deck of a bridge over canal had to be renewed. This was the Dinzer bridge in Friesland (the Netherlands). The old deck profiles were made of tropical hard wood, and in this case the new deck profiles were made from EoU thermoset composite. The profiles should resemble the original hard wood profiles by their dimensions (thickness 95 mm and width 245 mm) and their appearance and must be suitable for heavy traffic.

The profiles were designed with an outer layer of virgin glass fibre-reinforced polyester and a core build up from re-used EoU thermoset composite flakes embedded in virgin polyester resin. First prototypes of the profiles had to be mechanically tested to verify mechanical loading strength. Successful tests confirmed that the beam design fulfils the requirements for Dutch infra structural design [46]. The profiles that were made for the new deck, were produced using a method of resin casting in an open mould.

The profiles were mounted on a new bridge structure and give the appearance of a traditional deck made with wood deck profiles (see Figure 1.31). An important advantage of re-used EoU composite is that the service life is much longer, and maintenance is lower as compared with wood decking.

(a)   (b)

Figure 1.31: Deck profiles of re-used composite at the Dinzer bridge: (a) installation and (b) result (obtained by ten Busschen).

1.7 Mechanical recycling by grinding

Mechanical grinding produces small size fragments and particles to be used as reinforcement or filler materials. The geometry and composition of the input material is less critical compared to the processes described in the preceding sections, and as such a grinding process is capable of absorbing materials which do not fit or are a by-product (i.e. cutting losses) of the other recovery processes described in this chapter. For these materials, and for reused products that cannot go through another use cycle, mechanical grinding is a viable option.

1.7.1 Process outline

Mechanical grinding breaks down composite scrap into smaller pieces to roughly separate fibre and resin fractions. The energy demand of mechanical grinding lies between 0.1 and 4.8 MJ/kg, depending on the used machinery and process scale. Powering the motor of the granulator or hammer mill machine constitutes most of the energy demand. The pre- and post-recycling stages, such as shredding and sieving, are not as energy intensive as the actual downsizing process [48].

Grinding does not use any chemicals and, if done in a controlled environment, does not cause atmospheric or water pollution. In addition, no high-end technical and expensive equipment is needed compared to other recycling methods, it allows for processing of larger amounts of waste at higher throughputs [49]. However, the inferior quality of recovered recyclate compared to the virgin product complicates finding market applications for the recyclate. Economically, the recyclate is hardly competitive with virgin raw filler materials [4, 49].

1.7.2 Process

In the mechanical grinding process, there are three main steps to be distinguished: shredding, grinding, and classifying (Table 1.4). First, the material is shredded down to chunks of 50–100 mm using a slow speed cutting or crushing mill. Shredding makes it easier to remove metal inserts and, if done at the waste location, the volume reduction makes transport easier. Secondly, the chunks are ground down further into fragments of 10 mm–50 μm using a cutting or hammer mill (see Figure 1.32 [50]). Both have their own benefits: cutting mills give more homogeneous fibre length distribution and longer fibres, whereas the blades of the hammer mills do not require sharpening, thus reducing wear and increasing the output. There is no significant difference between the two concerning the resin content of the recyclate [51]. Thirdly, the fragments are separated on their contents and size using cyclones and sieves [50].

Table 1.4: The mechanical recycling process ([7, 50]).

	1. Shredding	2. Grinding/milling	3. Classifying
Process	Reduce waste materials into pieces	Grind pieces into fragments	Separating the fragments into resin-rich powders and fibrous fragments
Particle size[a]	50–100 mm	10 mm–50 μm	–
Equipment	Slow speed cutting or crushing mill	High-speed cutting or hammer mill	Cyclones and sieves

[a]At the end of the step.

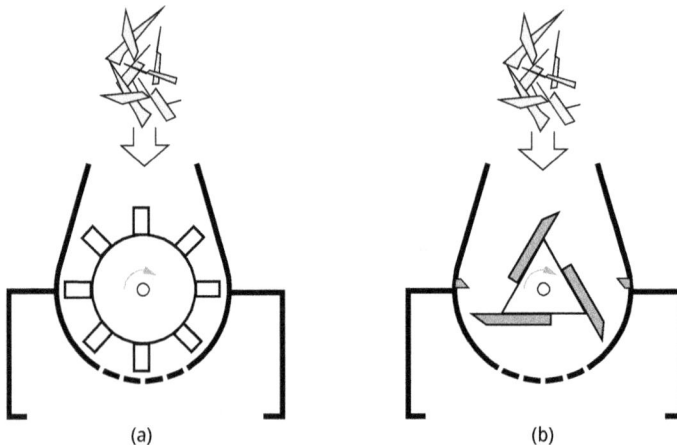

(a) (b)

Figure 1.32: simplified diagrams of a hammer mill (a) and cutting mill (b) (obtained by Joustra from [52]).

1.7.3 Outputs

It is important to note that there is no complete separation of the two materials; the fragments will almost always consist of mixed materials. The recyclate of the mechanical grinding are mixtures of resin, fibre, and filler. In this, two main fractions can be distinguished: fine and coarse. The finer fractions are powders with a higher resin and filler proportion compared to the original composite whereas the coarser fractions are more fibrous with higher fibre content [50].

Resin-rich powder can be used as a filler. The powder contains a significant proportion of low-density polymer and as such has a lower density than conventional fillers. This could lead to weight savings compared to using calcium carbonate filler. Even though filler substitute recyclates are more expensive by weight than traditional fillers, the lower density means they are cheaper per unit volume [50]. This could be favourable for industries pursuing weight and cost savings, such as the automotive industry. However, the economic viability of the filler materials is challenged by the low cost of primary fillers (e.g. calcium carbonate or silica). Also, the incorporation level as a filler is quite limited (<10 wt.%) because of the deterioration in mechanical properties and increasing processing problems at higher content due to the higher viscosity of the compound [7].

Fibrous fragments range from powders, fibre-particulate bundles, and fibre tows to woven platelets. Each is a combination of resin and reinforcements, complicating assessment of recycled fibre properties. The length of the fragments varies depending on initial fibre length, composite type [53], and scrap feeding size. Bream and Hornsby [53] report that the structural integrity of the fibres is preserved, with fibres up to 10 mm being retained and 69% of the recyclate particulates greater than 1 mm in diameter. The fibres produced by ERCOM Composite Recycling in Germany range from <0.25 mm to 3–20 mm particle size or fibre length [50].

The fibrous recyclate can be used as filler, as reinforcement for new composites in short-fibre applications such as bulk moulding compound (BMC), sheet moulding compound (SMC) parts, or as inclusions in injection-moulded products [54]. Dutch start-up Extreme Eco Solutions developed paving stones based on ground wind turbine blade composites (Figure 1.33) [55]. The tiles promise to be lower in weight and to contain up to 85% of ground composite as filler [56]. Using the recyclate as reinforcement represents a higher value route as the reinforcing value of the fibre is (partially) retained, whereas the filler replaces very low value materials, and more energy is needed to grind to fine filler [48].

In general, recyclate performance in composites is inferior to virgin fibres [50]. Even with low reincorporation rates, the resulting mechanical properties of the new composite material are significantly impaired due to a poor bonding between the recyclates and the new resin [7]. To improve bonding, additional treatments such as grafting or coupling agents [57] or increased mixing times are needed [7]. Finally, it is important that in addition to the remainder of the virgin fibres, additional longer fi-

Figure 1.33: Paving tile based on ground composite material [55] (with kind permission: taken from Extreme Eco Solutions, Langestraat 37, the Netherlands).

bres are added, as these will compensate for the deleterious effect of the recyclate [50].

1.8 Discussion

Composites present a challenge in achieving a circular economy because of their inherent materials mixture and complicated reprocessing. Mechanical reuse and recycling contribute to closing the loop by reprocessing these materials and making them available for reuse. The need for such processes is evident given the increasing amounts of EoU composite material becoming available, especially in the construction, wind energy, and maritime sectors. Following the estimated volumes of wind turbine blades, boat hulls, and feedstock silos, 3,500 ton of GFRP material would need to be processed in the Netherlands and over 135,000 ton in the EU. The first step in mechanical processing is usually reducing large products to transportable and processable segments, these are then subject to further mechanical treatments.

In mechanical processing of composites, we distinguish two main strategies: structural reuse and material recycling (Table 1.2). Structural reuse focuses on reusing material as-is with the fibre-resin combination and (part of) the laminate structure intact. Recycling processes aim for fibre-resin separation or processing into particle fractions. The presented mechanical reuse and recycling processes are complementary to each other and can be applied in successive product life cycles.

Structural reuse can be used to slow the flow of materials through the economic system by extending the material lifetimes and thus pushing back the moment of recycling. As such, structural reuse can keep the materials in use until processing capacity and market are developed. From a materials flow perspective, structural reuse could meet the basic requirements for a successful recovery system. A successful recovery system should balance supply, processing capacity, and market for the recyclate [58].

Additionally, a continuous flow of waste is required to set up and run a mechanical recycling plant and to produce products from the recyclates. The viability of such an operation depends on, and is often hampered by, the availability of EoU composite materials [59]. With the expected composite waste volume from for example end of life wind turbines, the required waste volume will not be hard to achieve. However, without a viable new end-application for the recyclate, going through the whole process might not be justifiable or even sustainable.

Reuse across contexts and extending material use beyond its original design lifetime brings additional challenges in terms of performance, safety, and certification. Materials in use are subject to wear, most notably mechanical fatigue, and environmental degradation such as ageing and surface erosion. Therefore, the quality of the material is not always exactly known when taken out of use. In addition, reuse in another context changes the loading conditions. This can be dealt with by not using the material to its primary full capacity.

To optimise reuse and warrant safety of structural elements in terms of mechanical performance, an improved assessment of residual material quality is desirable. Three approaches to assess residual mechanical performance of composite materials are identified. The first approach assumes the material is still within its original specification when recollected. In the second, the residual material quality is modelled based on original documentation and in-use monitoring, which can be facilitated by progress in modelling fatigue behaviour of composites [60, 61]. The third approach considers testing samples of the acquired material to experimentally determine residual quality [62]. The first would provide a rough estimate, the second a good indication, and the third approach an actual, albeit local, and time-consuming, measurement of residual properties. The desired level of accuracy will depend on the intended reuse case and required safety margins.

Mechanical reuse and recycling can be used to realise several consecutive use cycles. Figure 1.34 illustrates five cases of glass fibre composite use and reuse, building on the case of wind turbine blades. First, the composite material is used as a wind turbine blade with a design life of 25 years. Second, major blade parts can be repurposed as load carrying beams in a slow traffic bridge, adding approximately 60 years to the material life. In a third step, the structure can be segmented (resized) for reuse as construction elements in e.g. furniture with an estimated lifespan of 15 years. Fourth, the GFRP can be processed into oblong flakes and used in retaining walls for another 60 years. Finally, as a fifth option, the materials can still be mechanically recycled, reprocessed, and used as a filler. These successive use cycles extend the potential material lifetime from 25 to up to 200 years. In addition to extending the material lifetime, reuse and recycling avoid use of primary materials such as steel beams for construction or wood for retaining walls.

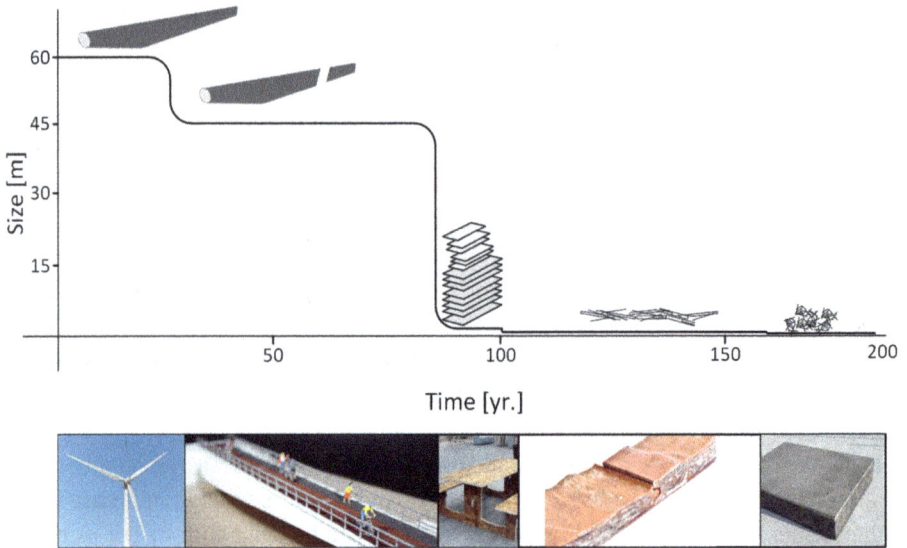

Figure 1.34: Structural reuse of wind turbine blades, extending lifetime and preserving value of the material (obtained by Joustra).

1.9 Conclusions

Mechanical reuse and recycling processes are attractive for composite products because of their capacity to preserve the material characteristics and prolong material lifetime at several levels of materials integrity. Repurposing and segmentation of large parts provides direct access to high performance materials at relatively low cost. Processing composites into flakes and embedding those in new products preserves the intrinsic strength properties. Grinding delivers filler material that could increase stiffness, while reducing cost and potentially weight for manufacturing new products. Thus, structural reuse, and to some extent recycling processes, retain the beneficial properties of composite materials and make it available for next use cycles.

The mechanical reuse and recycling processes described in this chapter are complimentary to each other. The material lifespan can effectively be extended by subsequently reusing parts, fragments and shreds. Next to such serial application of these recovery pathways, they may be used in parallel to ensure high reprocessing percentages. The offcuts and other "leftover" materials from repurposing or structural reuse provide immediate input for reprocessing into flakes or regrind. However, the general trend across these processes is a reduction in size and hence a reduction in material properties relative to the original use case.

The best planning of recovery actions will depend on the availability of resources, processing capacity, and potential market for the recyclate. Resource availability is starting to show, and the emerging return volumes are already reason for concern for some industry sectors. It is therefore necessary to get reprocessing capacity in place to deal with these increasing volumes of EoU composite materials. A key driver for development of recovery facilities is also the development of a market, or in other words, applications in which these recovered materials can effectively be reused. Development of such processes and materials can, in conjunction with product-level strategies and other recycling technologies, contribute to establishing a closed-loop for composite materials.

References

[1] Ellen MacArthur Foundation Towards the Circular Economy 1: Economic and Business Rationale for an Accelerated Transition Available online: https://www.ellenmacarthurfoundation.org/publications/towards-the-circular-economy-vol-1-an-economic-and-business-rationale-for-an-accelerated-transition.

[2] Korhonen, J.; Honkasalo, A.; Seppälä, J. Circular Economy: The Concept and Its Limitations. Ecol. Econ. 2018, 143, 37–46, doi:10.1016/J.ECOLECON.2017.06.041.

[3] Velenturf, A.P.M.; Purnell, P. Principles for a Sustainable Circular Economy. Sustain. Prod. Consum. 2021, 27, 1437–1457, doi:10.1016/j.spc.2021.02.018.

[4] Yang, Y.; Boom, R.; Irion, B.; Van heerden, D.J.; Kuiper, P.; De wit, H. Recycling of Composite Materials. Chem. Eng. Process. Process Intensif. 2012, 51, 53–68, doi:10.1016/j.cep.2011.09.007.

[5] Beauson, J.; Brøndsted, P. Wind Turbine Blades: An End of Life Perspective. In MARE-WINT: New Materials and Reliability in Offshore Wind Turbine Technology; Ostachowicz, W., McGugan, M., Schröder-Hinrichs, J.U., Luczak, M., Eds.; Cham (Switzerland): Springer OPEN., 2016; pp. 421–432 ISBN 9783319390956.

[6] Nijssen, R.P.L. Composite Materials; 1st ed.; Bussum the Netherlands: VKCN., 2015; Vol. 3; ISBN 978-90-77812-471.

[7] Oliveux, G.; Dandy, L.O.; Leeke, G.A. Current Status of Recycling of Fibre-reinforced Polymers: Review of Technologies, Reuse and Resulting Properties. Prog. Mater. Sci. 2015, 72, 61–99, doi:10.1016/j.pmatsci.2015.01.004.

[8] Mativenga, P.T.; Sultan, A.A.M.; Agwa-Ejon, J.; Mbohwa, C. Composites in a Circular Economy: A Study of United Kingdom and South Africa. Procedia CIRP 2017, 61, 691–696, doi:10.1016/j.procir.2016.11.270.

[9] Jensen, J.P.; Skelton, K. Wind Turbine Blade Recycling: Experiences, Challenges and Possibilities in a Circular Economy. Renew. Sustain. Energy Rev. 2018, 97, 165–176, doi:10.1016/j.rser.2018.08.041.

[10] Joustra, J.; Flipsen, B.; Balkenende, R. Structural Reuse of Wind Turbine Blades through Segmentation. Compos. Part C 2021, 5, doi:10.1016/j.jcomc.2021.100137.

[11] Asmatulu, E.; Twomey, J.; Overcash, M. Recycling of Fiber-Reinforced Composites and Direct Structural Composite Recycling Concept. J. Compos. Mater. 2014, 48, doi:10.1177/0021998313476325.

[12] Ten busschen, A. Industrial Re-Use of Composites. Reinf. Plast. 2020, 64, 155–160, doi:10.1016/j.repl.2020.04.073.

[13] Schreuder, P.; Hermans, K.W.; Ten busschen, A. Hergebruik van Thermoharde Composieten – Onderzoek Naar Verwerken van End-of-Life Producten tot Versterkingselementen.; Zwolle: Christelijke Hogeschool Windesheim, 2017;

[14] Ten busschen, A.; Bouwmeester, Schreuder, P. Hergebruik van Thermoharde Composieten - Onderzoek van Afvalproductsoorten; Zwolle: Christelijke Hogeschool Windesheim, 2016;

[15] Liu, P.; Barlow, C.Y. Wind Turbine Blade Waste in 2050. Waste Manag. 2017, 62, 229–240, doi:10.1016/j.wasman.2017.02.007.

[16] ICF; Deloitte; South East Marine; Sea Teach; IEEP Assessment of the Impact of Business Development Improvements around Nautical Tourism; EUROPEAN COMMISSION Directorate-General: Brussels, 2016; ISBN 978–92-79-67732-8.

[17] IMO; Office for the London Convention/Protocol and Ocean Affairs End-Of-Life Management of Fibre-reinforced Plastic Vessels: Alternatives to At Sea Disposal; 2019;

[18] Riedewald, F.; Sousa-gallagher, M. Novel Waste Printed Circuit Board Recycling Process with Molten Salt. MethodsX. 2015, 2, 100–106.

[19] Ghosh, B.; Ghosh, M.K.; Parhi, P.; Mukherjee, P.S.; Mishra, B.K. Waste Printed Circuit Boards Recycling: An Extensive Assessment of Current Status. J. Clean. Prod. 2015, 94, 5–19, doi:10.1016/j.jclepro.2015.02.024.

[20] Lantz, E.; Wiser, R.; Hand, M. The Past and Future Cost of Wind Energy; Presented at the 2012 World Renewable Energy Forum, Denver, Colorado May 13–17, 2012, 2011.

[21] Memija, A. 123-Metre Wind Turbine Blade Rolls Out in China Available online: https://www.offshore wind.biz/2022/09/07/123-metre-wind-turbine-blade-rolls-out-in-china/ (accessed on 12 December 2022).

[22] Resor, B.R. Definition of a 5MW/61.5 m Wind Turbine Blade Reference Model. Albuquerque, New Mex. USA, Sandia Natl. Lab. SAND2013-2569 2013 2013, 50.

[23] Mishnaevsky, L.; Branner, K.; Petersen, H.N.; Beauson, J.; McGugan, M.; Sørensen, B.F. Materials for Wind Turbine Blades: An Overview. Materials (Basel). 2017, 10, 1–24, doi:10.3390/ma10111285.

[24] Joustra, J.; Flipsen, B.; Balkenende, R. Structural Reuse of High End Composite Products: A Design Case Study on Wind Turbine Blades. Resour. Conserv. Recycl. 2021, doi:10.1016/j.resconrec.2020.105393.

[25] Bank, L.C.; Arias, F.R.; Yazdanbakhsh, A.; Gentry, T.R.; Al-Haddad, T.; Chen, J.F.; Morrow, R. Concepts for Reusing Composite Materials from Decommissioned Wind Turbine Blades in Affordable Housing. Recycling 2018, 3, doi:10.3390/recycling3010003.

[26] Vestas Product Disposal Specifications; 2022;

[27] LM Windpower Innovation Is the Root of the Future Available online: https://www.lmwindpower. com/en/products-and-services/we-know-blades/innovation-is-the-root-of-the-future (accessed on 8 December 2022).

[28] Fu, R. Covestro and TMT Launch their 1000th Polyurethane Wind Rotor Blade, Leverkusen/Shanghai: Covestro AG, 2022, pp. 1–3.

[29] Germanischer Lloyd Rules and Guidelines Industrial Services: Guideline for the Certification of Offshore Wind Germanischer Lloyd Eds.; Germanischer Lloyd, Hamburg, 2010;

[30] Tazi, N.; Kim, J.; Bouzidi, Y.; Chatelet, E.; Liu, G. Waste and Material Flow Analysis in the End-of-Life Wind Energy System. Resour. Conserv. Recycl. 2019, 145, 199–207, doi:10.1016/j.resconrec.2019.02.039.

[31] WindEurope Repowering and Lifetime Extension: Making the Most of Europe's Wind Energy Resource; 2017;

[32] Ten busschen, A. Revolutionary re-use of Polyester Boats. The Report. Issue 81, September 2017, pp. 50–55.

[33] Speksnijder, S. Bridge of Blades Available online: http://www.stijnspeksnijder.com/gallery/bridge-of-blades/ (accessed on 16 April 2020).

[34] Nagle, A.J.; Ruane, K.; Gentry, T.R.; Bank, L.C.; Dunphy, N.; Mullally, G.; Paul, G. Life Cycle Sustainability Assessment of a Pedestrian Bridge Made from Repurposed Wind Turbine Blades. 2022, 443–448.

[35] Bank, L.C.; Gentry, T.R.; Al-Haddad, T.; Alshannaq, A.; Zhang, Z.; Bermek, M.; Henao, Y.; McDonald, A.; Li, S.; Poff, A.; et al. Case Studies of Repurposing FRP Wind Blades for Second-Life New Infrastructure. Curr. Perspect. New Dir. Mech. Model. Des. Struct. Syst. 2022, 2019, 501–502, doi:10.1201/9781003348450-235.

[36] Blade Made Sound Barrier Available online: https://blade-made.com/portfolio-items/blade-barrier/.

[37] Speksnijder Reuse of Wind Turbine Blades in a Slow Traffic Bridge, Delft University of Technology: Delft, 2018.

[38] Costa, S.; André, A.; Mattson, C.; Bru, T.; Ghafoor, A. Mapping and Digitization of the Flow of Wind Turbine Blades Goal & Agenda. In Proceedings of the International Conference on Sustainable Wind Turbine Blades: New Materials, Recycling and Future Perspectives; Beauson, J., Mishnaevsky, L., Eds.; Roskilde, 2022.

[39] Stone, M. Engineers Are Building Bridges with Recycled Wind Turbine Blades Available online: https://www.theverge.com/2022/2/11/22929059/recycled-wind-turbine-blade-bridges-world-first (accessed on 24 January 2023).

[40] Pronk, S. Repurposing Wind Turbine Blades as a Construction Material, Delft: Delft University of Technology, 2022.

[41] Kiss, P.; Stadlbauer, W.; Burgstaller, C.; Stadler, H.; Fehringer, S.; Haeuserer, F.; Archodoulaki, V.M. In-House Recycling of Carbon- and Glass Fibre-Reinforced Thermoplastic Composite Laminate Waste into High-Performance Sheet Materials. Compos. Part A Appl. Sci. Manuf. 2020, 139, 106110, doi:10.1016/j.compositesa.2020.106110.

[42] Cousins, D.S.; Suzuki, Y.; Murray, R.E.; Samaniuk, J.R.; Stebner, A.P. Recycling Glass Fiber Thermoplastic Composites from Wind Turbine Blades. J. Clean. Prod. 2019, doi:10.1016/j.jclepro.2018.10.286.

[43] Medici, P.; Van den dobbelsteen, A.; Peck, D. Safety and Health Concerns for the Users of a Playground, Built with Reused Rotor Blades from a Dismantled Wind Turbine. Sustain. 2020, 12, 1–25, doi:10.3390/su12093626.

[44] Cox, H.L. The Elasticity and Strength of Paper and Other Fibrous Materials. Br. J. Appl. Phys. 1952, 3, 72–79, doi:10.1088/0508-3443/3/3/302.

[45] Ning, H.; Lu, N.; Hassen, A.A.; Chawla, K.; Selim, M.; Pillay, S. A Review of Long Fibre Thermoplastic (LFT) Composites. Int. Mater. Rev. 2020, 65, 164–188, doi:10.1080/09506608.2019.1585004.

[46] Ten busschen, A. Industrial Re-Use of Composites. In Waste Material Recycling in the Circular Economy; Achilias, D.S., Ed.; Intech Open: London, 2021.

[47] Halliwell, S. Fibre Reinforced Polymers in Construction: Durability; 2003;

[48] Job, S.; Leeke, G.; Mativenga, P.T.; Oliveux, G.; Pickering, S.J.; Shuaib, N.A. Composites Recycling: Where Are We Now?; Berkhamsted, United Kingdom: CompositesUK, 2016;

[49] Ribeiro, M.; Fiúza, A.; Ferreira, A.; Dinis, M.; Meira Castro, A.; Meixedo, J.; Alvim, M. Recycling Approach towards Sustainability Advance of Composite Materials' Industry Recycling 2016, 1, 178–193, doi:10.3390/recycling1010178.

[50] Pickering, S.J. Recycling Technologies for Thermoset Composite Materials – Current Status. Adv. Polym. Compos. Struct. Appl. Constr. ACIC 2004 2005, 37, 1206–1215, doi:10.1016/B978-1-85573-736-5.50044-3.

[51] Schinner, G.; Brandt, J.; Richter, H. Recycling Carbon-Fibre-Reinforced Thermoplastic Composites. J. Thermoplast. Compos. Mater. 1996, 9, 239–245, doi:10.1177/2F089270579600900302.

[52] Macko, M. Size Reduction by Grinding as an Important Stage in Recycling. Post-Consumer Waste Recycl. Optim. Prod. 2012, 273–294, doi:10.5772/33969.

[53] Bream, C.E.; Hornsby, P.R. Comminuted Thermoset Recyclate as a Reinforcing Filler for Thermoplastics – Part I Characterisation of Recyclate Feedstocks. J. Mater. Sci. 2001, 36, 2977–2990.

[54] Mativenga, P.T.; Shuaib, N.A.; Howarth, J.; Pestalozzi, F.; Woidasky, J. High Voltage Fragmentation and Mechanical Recycling of Glass Fibre Thermoset Composite. CIRP Ann. – Manuf. Technol. 2016, 65, 45–48, doi:10.1016/j.cirp.2016.04.107.

[55] Extreme Eco Solutions Re-Tile : Circulaire Sierbestrating En Wandtegels Available online: https://extreme-ecosolutions.com/ (accessed on 24 January 2023).

[56] IJsenbrand, B. Extreme Eco Solutions Recyclet Oude Windmolens | Voor de Wereld van Morgen Available online: https://www.voordewereldvanmorgen.nl/artikelen/extreme-eco-solutions-recyclet-oude-windmolens (accessed on 24 January 2023).

[57] Bream, C.E.; Hornsby, P.R. Comminuted Thermoset Recyclate as a Reinforcing Filler for Thermoplastics – Part II Structure-Property Effects in Polypropylene Compositions. J. Mater. Sci. 2001, 36, 2965–2975, doi:10.1023/A:1017962722495.

[58] Pickering, S.J. Recycling Technologies For Thermoset Composite Materials. Compos. Part A 2006, 37, 1206–1215, doi:10.1016/B978-1-85573-736-5.50044-3.

[59] Vijay, N.; Rajkumara, V.; Bhattacharjee, P. Assessment of Composite Waste Disposal in Aerospace Industries. Procedia Environ. Sci. 2016, 35, 563–570, doi:10.1016/j.proenv.2016.07.041.

[60] Rubiella, C.; Hessabi, C.A.; Fallah, A.S. State of the Art in Fatigue Modelling of Composite Wind Turbine Blades. Int. J. Fatigue 2018, 117, 230–245, doi:10.1016/j.ijfatigue.2018.07.031.

[61] Nijssen, R.P.L.; Brøndsted, P. Fatigue as a Design Driver for Composite Wind Turbine Blades. In Advances in Wind Turbine Blade Design and Materials; Brøndsted, P., Nijssen, R.P.L., Eds; Cambridge: Woodhead Publishing, 2013; pp. 175–209 ISBN 9780857094261.

[62] Alshannaq, A.; Scott, D.; Bank, L.C.; Bermek, M.; Gentry, R. Structural Re-Use of de-Commissioned Wind Turbine Blades in Civil Engineering Applications. Proc. Am. Soc. Compos. – 34th Tech. Conf. ASC 2019 2019, https://dpi-proceedings.com/index.php/asc34/.

Matthew Keith, Gary Leeke

2 Chemical recycling of polymer matrix composites

2.1 Introduction

At the simplest level, fibre-reinforced polymers (FRPs) consist of a fibre reinforcement (such as carbon, glass, aramid, or natural fibres) held together by a polymer matrix. Globally, ~90–95% of all polymeric composites are glass fibre-reinforced polymers (GFRPs), with the bulk of the remainder consisting of carbon fibre-reinforced polymers (CFRPs) [1]. As such, the focus within this chapter is on the chemical recycling of GFRPs and CFRPs. To retain the maximum value of these materials, their recycling necessitates the separation of the fibres from the polymer matrix. Often, glass fibres hold a lower economic value than the resin, while carbon fibres are significantly more valuable. For this reason, the recycling of CFRPs has prioritised the recovery of the fibres while GFRP recycling mostly focuses on organic products. Unfortunately, however, the bulk of GFRP waste is currently disposed of in landfills, with some used in cement kilns [1], while degraded resin from CFRPs typically only finds use as a fuel source [2]. Both landfilling and incineration are widely recognised as unsustainable end-of-life strategies; however, before chemical recycling strategies for composites can be discussed, it is first necessary to understand what these materials are.

2.1.1 Common polymer matrices

In the manufacturing of polymer composites, there are two broad classes of polymer which are commonly used: either thermoplastics or thermosets. The former category consists of long polymer chains held together only by weak intermolecular van der Waals forces. As such, these polymer chains become more mobile when heated, with the bulk plastic softening until it eventually melts. A wide range of thermoplastic matrix materials exists with a wide range of glass transition and melting temperatures. Amongst the lowest of these are the common consumer plastics polyethylene and polypropylene, which are more typically used in GFRPs for relatively low-performance applications. At the opposite end of the spectrum, "engineering plastics" with much higher thermal durability, strength, and stiffness are used in more demanding areas, such as structural components of aircraft, vehicles, and wind turbines. These include polyamides (PAs), polycarbonate (PC), polyphenylene sulphide (PPS), and polyetherether ketone (PEEK) [3]. Thermosets, meanwhile, have an extra layer of complexity. The polymer chains tend to be shorter than those found in thermoplastics; however, these chains are also held together by covalently bonded cross-linking agents, or hardeners.

https://doi.org/10.1515/9783110754438-002

During manufacture, the prepolymer (e.g., bisphenol A diglycidyl ether, BADGE) is mixed with a hardener (such as diamino-diphenyl-sulphone, DDS), heated, and possibly pressurised. This heating induces a curing reaction where the prepolymer chains become bonded together. The resulting thermoset resin then does not soften or melt when heated, although they will thermally degrade at temperatures above 450 °C [4].

A third component frequently, if not always, used in composites is referred to as sizing. This is a generic name given to the polymeric coating applied to the surface of the fibres during their manufacture. The purpose of this is to protect the filaments and improve adhesion between the fibre and the matrix. Like the different polymers, there is also a wide variety of different sizing formulations used. Fortunately, the sizing generally makes up only a small proportion of the mass of the final product and rarely exceeds 3 wt.% of the resin [5]. Above this threshold, reductions in mechanical performance have been seen; for every percentage increase in the mass of sizing, the interfacial shear strength decreases by 2–3% [6]. In addition to sizing, fillers, coatings, and other additives which give the final composite additional properties are becoming more common. For example, ammonium polyphosphate may be used as a flame retardant [7], while silica or rubber nanoparticles may be included to reduce chip damage, thus improving the machining of the composite [8]. In the latter example, the filler materials accounted for up to 20 wt.% of the matrix material; this significant proportion will, therefore, need to be considered upon recycling the composite.

Due to the wide variety of different polymer matrices, sizing materials, and fillers, the separation of composites prior to their recycling is still a significant barrier to achieving a fully circular economy. Recent advances in hyperspectral imaging (HSI) do make it possible to identify different polymers, but currently, this has only been applied to the inspection of CFRPs at a research stage [9, 10]. In the future, it may be possible to use HSI, alongside automated systems and artificial intelligence, to separate composites based on matrix material. At present, however, it seems likely that any recycling process will need to handle a mixed feedstock and the subsequent generation of a complex mixture of organic compounds due to the decomposition of the resin. This chapter outlines the general principle of chemical recycling and briefly considers the economic and environmental reasons for developing such technologies. Recent advances in recycling processes are presented and categorised according to polymer type and dominant degradation mechanism, before focusing on the properties of the recovered fibres. The chapter then considers the influence of additives on recycling systems and summarises the available information regarding the current technological readiness of these systems, as well as their environmental impacts. Finally, this chapter concludes with an assessment of current challenges and barriers to delivering a circular economy for composites, along with a set of recommendations for how these might be addressed.

2.1.2 Principle of chemical recycling

Unlike thermolytic processes, such as pyrolysis, chemical recycling uses a chemical agent to either depolymerise or degrade the matrix of an FRP. Depolymerisation refers to the specific reaction which liberates the original monomers used in the polymer, while degradation refers to the non-specific decomposition of the matrix. This likely yields a mixture of different organic compounds which, to be re-used in high-value applications, would almost certainly need to be upgraded. The term "chemical recycling" is often used interchangeably with "solvolysis", which refers to the chemical bonds broken by the action of a solvent.

During the chemical recycling process, a solvent or other reactant diffuses into the polymer, causing it to swell and, hence, push apart the layers of multi-ply FRPs. Chemical bonds are broken, either by the solvent or a catalyst, resulting in the production of smaller molecules, which diffuse out of the composite and into the bulk fluid. Figure 2.1 shows samples of a CFRP, expanded layers of carbon fibre due to solvent-induced swelling, penetration of the solvent to the centre of the composite, and, finally, clean carbon fibre tows. Common solvents include water, short-chain alcohols, acetone, ethylene glycol (EG), phenol, and amines, which are either supplied alone or occasionally in combination with water. Common bases include potassium hydroxide or sodium hydroxide, while nitric, sulphuric, or acetic acids are frequently investigated as acidic media. Hydrogen peroxide has also been widely considered as an oxidising agent, while other ad-

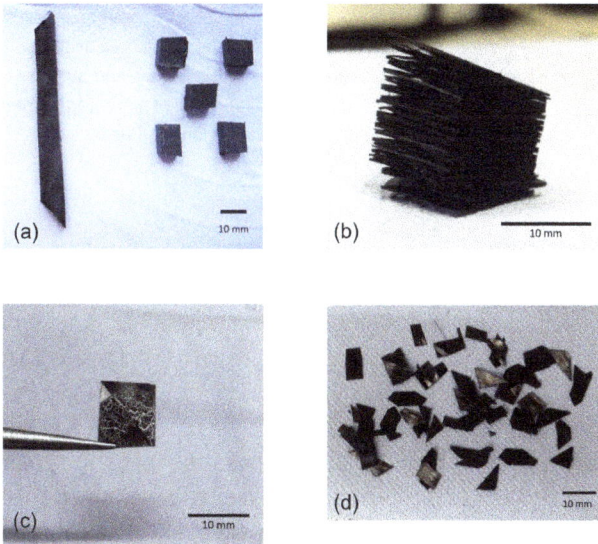

Figure 2.1: Samples of a CFRP: (a) before processing, (b) swollen after partial degradation of the polymer, (c) penetration of the solvent to the centre of the sample, and (d) clean carbon fibre tows with the resin completely decomposed. Images taken from [11].

ditives include potassium phosphate and metal chlorides. The choice of solvent, catalyst, or additive depends largely on the polymer matrix; however, as mentioned previously, it is likely that any commercial recycling process will need to be able to handle a mixture of polymer composite waste, which will, hence, generate a complex mixture of recyclate.

Although chemical recycling of FRPs is generally considered to be at a lower technology readiness level (TRL) than pyrolysis, it holds a number of key advantages. The lower temperature required often prevents any residual char from forming on the surface of the fibres. Glass fibres, in particular, are highly thermally sensitive, and so, to preserve the mechanical properties of the reinforcement agent, separation from the matrix under mild conditions is desirable. Within pyrolysis, it is not considered practical to recover any of the organic products from the matrix, meaning that gaseous products from the resin are typically oxidised and vented, or used as fuel in the process. Chemical recycling offers an opportunity to improve resource efficiency through the recovery of potentially useful organic compounds. Samples of this liquid product, which can be obtained from CFRP recycling, are shown in Figure 2.2.

Figure 2.2: Samples of the organic liquid products obtained after the solvolysis of a commercial carbon fibre-reinforced epoxy using an acetone/water mixture at 320 °C for 120, 60, and 30 min. The darker colour is indicative of a greater degree of resin decomposition at longer times.

2.1.3 Low and high temperature processes

Historically, chemical recycling processes have been broadly categorised as either low temperature and pressure (LTP) or high temperature and pressure (HTP) [12, 13]. LTP conditions were identified as less than 200 °C and atmospheric pressure. Due to these relatively mild conditions, the energy demand, and subsequent running costs, and environmental impact of a commercial recycling process may be significantly less than those of an HTP system. As it is not necessary to contain particularly high pressures, processing equipment, such as reactors, is also likely to be cheaper, resulting in a lower capital cost of the plant. The lower temperatures also offer better reaction control; side reactions and further decomposition of organic molecules are less likely to occur. This leads to a higher recovery of the decomposed polymer, though not always the curing agent [4]. However, strong acids, alkalis, oxidising agents, or other additives are often needed to fully break down the polymer. It may also be necessary to use a stirred reactor to enhance the mass transfer of the reactants. This is likely to also entangle the fibres either with each other or around the impeller, which leads to more down-time and labour-intensive, manual separation of the fibres, and stirring equipment. As aligned fibres give a higher quality FRP than non-aligned fibres, it may also be necessary to straighten individual fibres or tows by hand (to some extent) to maximise the value of the recycled composite [14]. A carbon fibre alignment process, entitled HiPerDif, has been developed at the University of Bristol, with the spinout company Lineat demonstrating commercialisation potential [15]. A second disadvantage of LTP processes is that the use of strong acids or bases can be dangerous in terms of human health and safety of the environment and may lead to mixtures which are difficult to recycle or otherwise dispose of.

The use of high temperatures, typically above 250 °C, eliminates the need for these potentially harmful and dangerous chemicals necessary in LTP conditions. Higher temperatures also lead to faster reaction rates due to the exothermic nature of the process; hence, FRPs can be chemically recycled over time periods of a few minutes to a couple of hours, rather than 12 or more hours. Many HTP chemical recycling processes of FRPs involve the use of supercritical fluids. By increasing the temperature and pressure of a substance, it will move through the normal states of matter of solid to liquid to gas. A further increase causes the substance to transition into the supercritical state. Above the critical point, there is no phase boundary between the liquid and gaseous states, and as such, the SCF exhibits properties in between those of a liquid and a gas. They typically exhibit a high diffusivity, enhanced mass transfer, very low viscosity, and a pressure-dependent solvating power. As such, the solvent properties and, therefore, the reaction rates and selectivity, can be controlled by changing the pressure within a reactor. Unfortunately, the high pressures required necessitate expensive equipment, while the generation of high temperatures is likely to give a high energy cost. Both of these factors have thus far acted as significant barriers to the commercialisation of this technology, and so all chemical recycling of FRPs using HTP processes have remained in the research phase. As a result, there has been a drive in recent years to reduce temperatures and pressures through the adoption of solvent mixtures, green solvents such as ionic liquids, and novel catalysts. There are now reports of numerous systems which operate in the 200–250 °C temperature range, and so the previous categorisation of LTP and HTP systems is now quite limited. For this reason, this chapter utilises a new approach to characterise recycling processes, which is based on the polymer matrix, solvent system, and dominant reaction mechanism, as discussed in Sections 2.2 and 2.3.

2.2 Chemical recycling of fibre-reinforced thermoplastics

As previously mentioned, thermoplastics consist of long polymer chains, typically in an amorphous or semi-crystalline structure. Only weak intermolecular forces exist between the polymer chains, which grow weaker as the material is heated. This causes it to soften and, eventually, melt into a viscous liquid. These materials are inherently recyclable; the same chemical bonds formed during polymerisation can be broken, thus liberating the original monomers. It is possible to obtain a range of different products from the matrix of fibre-reinforced thermoplastics; however, most work has focused on recovering clean fibres rather than analysing the monomer mixture. In addition, the use of strong acids and bases has mostly been applied to thermoset-based composites. For these reasons, this section also discusses the recovery of monomers from the recycling of non-reinforced thermoplastics, where equivalent research has not yet been conducted with FRPs [16]. Although most LTP processes rely on the use of strong acids, alkalis, or oxidising agents, their use depends heavily on the type of polymer matrix. For high-performance thermoplastics, such as PEEK, there are no reports of being able to use LTP processes. Although thermoplastics can be thermally reformed, the majority of thermoplastics can be solubilised (and thus liberate their fibre reinforcements) using aggressive acids, bases, or oxidising agents, often in aqueous media and occasionally assisted with microwaves or ultrasound, as discussed in the following sections.

2.2.1 Hydrolysis

Broadly, hydrolysis reactions refer to molecular bonds being broken due to the action of water. In FRP matrices, water molecules interact with vulnerable chemical bonds within the polymer chain. These are typically esters (R-COO-R') or amides (R-CON-R'), which, when broken, result in the degradation of the polymer into smaller oligomers and hence reduce the overall molecular weight. A wide range of thermoplastics may be hydrolysed, although the conditions required are strongly dependent on the polymer itself. Neutral hydrolysis (i.e., without acidic or basic media) is possible; for example, the common consumer plastic polyethylene terephthalate (PET) may be degraded by water alone, but at least sub-critical conditions with temperatures and pressures in excess of 200 °C and 1.4 MPa are required [17]. Here, protons (H^+) or hydroxyl ions (OH^-) act as catalysts to primarily yield EG and terephthalic acid (TPA), alongside any fibre reinforcement. In one study, the TPA yield was 90% and recoverable at a purity of 97% when processed at 300 °C [18]. As both TPA and EG are monomers used in PET production, this technology represents potential for closed-loop applications; however, despite being extensively researched at the lab scale, PET hydrolysis has yet to be developed as a commercially feasible route for both the recycling of FRPs and consumer plastics [19].

Fibre-reinforced polyamides are amongst the most widely studied materials with regard to their hydrolysis. This matrix may be hydrolysed at either sub- or supercritical conditions to yield the original monomer, ε-caprolactam. It should, however, be noted that longer reaction times, coupled with higher process temperatures, can also lead to additional reactions when using water. For example, at 500 °C, only ε-caprolactam was generated after a reaction time of 10 min, while a range of smaller and heavier molecules were present at longer reaction times due to further reactions of the organic products. The smaller molecules were released as an unidentified gas, while heavier compounds included 1,8-diazacyclotetradecane-2,9-dione, formed due to the reaction of two ε-caprolactam molecules. Additional repolymerisation was also reported to take place at this temperature, which may leave a residue on the surface of the recovered fibres [20]. To avoid these side reactions, sub- or supercritical conditions in the range of 200–400 °C hence appears to be a suitable method for the recycling of fibre-reinforced polyamides; however, it should be noted that glass fibres, in particular, are thermally sensitive, and such processes may irreversibly damage this reinforcement, as discussed in Carbon fibres, however, may withstand these and, due to the generation of the original ε-caprolactam monomer, there is the possibility of realising an improved circular economy. A summary of selected polymer matrices, process conditions, and key organic products is provided in Table 2.1.

Table 2.1: Selected conditions for the neutral hydrolysis of common matrix materials and their primary degradation products (adapted from [13]).

Polymer matrix	Temperature (°C)	Pressure (MPa)	Reaction time (min)	Key organic products	Reference
PET	200–250	1.4–2.0	180–300	Terephthalic acid Ethylene glycol	[21]
PET	250–400	5.0–24.0	1–30	Terephthalic acid Benzoic acid 1,4-dioxane Acetaldehyde Isophthalic acid	[22]
PA6	280 350 400 500	Not given Not given Not given Not given	60 40 15 10	ε-Caprolactam ε-Aminocaproic acid	[23]

Table 2.1 (continued)

Polymer matrix	Temperature (°C)	Pressure (MPa)	Reaction time (min)	Key organic products	Reference
PA6	300	20	60	ε-Caprolactam	[24]
	330	25	60	ε-Minocaproic acid	
	360	30	10		
	400	35	5		
PA6	345	9	45	ε-Caprolactam	[25]
	370	14	45	ε-Aminocaproic acid	

To accelerate degradation reactions and/or minimise the temperature required, acids, bases, or oxidising agents may be supplied, which immediately provide the protons and hydroxyl ions needed for depolymerisation. Examples of these techniques are discussed in the following sections.

2.2.1.1 Acidic media

In the acidic hydrolysis of FRPs, the most investigated matrix appears to be polyamide. Typically, strong acids such as hydrochloric, sulphuric and formic acids are used. For example, Češarek et al. investigated the microwave-assisted recycling of a range of PAs, both alone and reinforced with carbon and glass fibres. Depolymerisation was achieved at 200 °C within a short reaction time of 10 min. However, long-chain PAs, and those with fibre reinforcement, need double the concentration of HCl to achieve the same degree of depolymerisation in the same reaction time. The authors suggest that the presence of glass fibres inhibits the reaction due to a reduction in the solvation of PA. Similar findings were reported due to the presence of carbon fibres and it was noted that higher reinforcement loadings also led to longer reaction times. Following the reaction, it was possible to recover high yields of the original monomers (adipic acid, hexamethylene diamine, 11-amino-undecanoic acid hydrochloride, 12-dodecanoic acid hydrochloride, and sebacic acid), although the products obtained differed depending on the initial PA [26].

In other research, acid solutions were gently warmed and refluxed with PA samples over a period of up to 20 h. The later addition of water resulted in the precipitation of low molecular weight fractions of degraded polymer. In a formic acid treatment, complete depolymerisation to the monomer alkyl cyanoacrylate (ACA) was not observed. However, refluxing with 30 vol.% HCl for 4 h was able to convert 93% of the PA6 into ACA. Upon increasing both the reaction time and concentration of HCl, there was minimal change to the yield of the monomers. This was attributed to potential repolymerisation occurring at the temperatures investigated. Sulphuric acid was less effective than HCl; however, a concentration of only 15 vol.% was used [27].

There has also been a little work investigating weak organic acids (acetic, propanoic, and butanoic) as catalysts in the hydrolysis of polyamides. Surprisingly, these weak organic acids were reported to double the rate of hydrolysis of the polyamide compared to a dilute HCl solution. All acids were supplied at the same pH, using temperatures in the range of 100–120 °C. It was found that butanoic acid resulted in the greatest rate of hydrolysis, as the longer carbon chain showed greater solubility into the polyamide. The authors also state that it is this higher solubility in the polyamide (in their undissociated state) which enables greater diffusion into the polymer, and hence greater degradation than what can be achieved with HCl [28].

As these examples have shown, both strong and weak organic acids are able to accelerate the depolymerisation of thermoplastics under low-temperature conditions. The following section discusses the effectiveness of basic media on the decomposition of selected thermoplastics.

2.2.1.2 Basic media

In addition to acidic media, alkaline hydrolysis of thermoplastics is also possible. This technology has been applied to a wide range of consumer plastics, such as PET, polypropylene, polyethylene, and polylactic acid, which are also often used in FRPs. Similarly, basic media may be used to depolymerise engineering thermoplastics, such as PEEK, although their much higher glass transition temperatures necessitate the use of higher processing temperatures, often above 300 °C.

Of the lower-performance thermoplastics, the hydrolysis of PET appears to be the most widely investigated. Early research demonstrated that the hydrolysis of PET flakes using sodium hydroxide (NaOH) was possible in the temperature range of 120–200 °C. This process yielded the disodium TPA salt, which may be subsequently acidified to recover TPA at a purity of 99.6%, making it suitable for direct polymerisation back to PET [29]. Phase transfer catalysts, which typically consist of an ionic hydrophilic head and a hydrophobic tail, have been considered. These enable immiscible reactants, such as water and a polymer, to react by transporting molecules across phase boundaries. For example, a quaternary ammonium salt enabled close to 100% recovery of TPA within 60 min when used in conjunction with a 10% NaOH solution and heated via microwaves [30]. Ultrasound is also able to accelerate depolymerisation reactions; in other work, the use of ultrasound reduced the reaction time for a 10% NaOH/tetrabutyl ammonium iodide aqueous system to depolymerise PET by about 30% [31]. More recently, "super basic" reagents have been investigated as one approach to accelerate the hydrolytic degradation of PET. For example, 1,8-diazabicyclo[5.4.0]undec-7-ene (DBU) acts to significantly raise pH levels to above 13. Although similar temperatures in the range of

120–150 °C were necessary, much faster reaction times of less than 60 min were sufficient to fully solubilise PET using a concentration of 1.5 M DBU [32].

For the recycling of high-performance materials, such as carbon fibre-reinforced PEEK, high temperatures of 300–400 °C, a co-solvent such as ethanol, propanol, or acetone, and a basic catalyst are all essential to break the C–O bonds. The strength of this polymer network is derived from its aromatic structure, which is illustrated by Figure 2.3 Despite the presence of an organic solvent, it is still thought that the dominant reaction mechanism here is hydrolysis; in the absence of water, the reaction proceeds slowly, and minimal concentrations of the major degradation product, phenol, are produced [33]. The role of the alcohol or acetone as a co-solvent may be to lower the critical point of the mixture and, thus, the reaction system benefits from the properties of a supercritical fluid without needing to reach the critical point of water. For water, this is 374 °C and 22 MPa [34], yet complete PEEK depolymerisation was achieved at 350 °C in just 30 min [33]. In addition to phenol, major degradation products include various phenolic derivatives, for example, 4-phenoxyphenol, 1-methyl-3-(4-methylphenoxy)benzene, and dibenzo furan [33]. It therefore does not seem possible to recover the original PEEK precursors (commonly 4,4'-difluorobenzophenone and hydroquinone [35]) following this recycling method. However, phenols (and its derivatives) are common platform chemicals, so it may be possible to recover some additional value for the organic residue recovered from such a system.

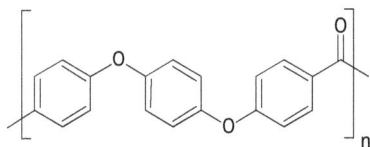

Figure 2.3: Chemical structure of PEEK [36].

As alluded to in the previous paragraph, there exist multiple reaction systems capable of solubilising thermoplastics, with hydrolysis representing just one such mechanism. Although alcohols are larger molecules than water, they are still capable of penetrating a polymer network, breaking down chemical bonds, and hence liberating clean fibres, as discussed in the following section.

2.2.2 Alcoholysis

Unlike hydrolysis, alcoholysis is a transesterification process in which polymer chains are cleaved through reaction with an alcohol, typically no larger than butanol (which contains four carbon atoms). This may occur in the presence of a catalyst or rely on the action of heat, pressure, and time to break vulnerable bonds within the polymer chain. This yields smaller oligomers or monomer derivatives, such as dimethyl terephthalate (DMT) from PET or methyl lactate from PLA. For fibre-reinforced thermoplastics, alcoholysis has some advantages over hydrolysis as it enables the use of milder conditions whilst facilitating the recovery of both high-value organics and intact fibres.

The depolymerisation of PET, either as a neat polymer or reinforced with glass fibres, has been widely studied. Methanolysis of PET in the presence of basic catalysts such as KOH or NaOH yields TPA and EG or DMT, both of which are useful precursors for closed-loop PET synthesis. Reaction times of less than 10 min at temperatures between 110 and 150 °C are often sufficient for complete depolymerisation [37]. However, the presence of fibres influences the reaction kinetics. For example, pure PET pellets underwent complete depolymerisation in just 1 min at 120 °C, whereas PET reinforced with glass fibres required 5 min under otherwise identical conditions [37]. This demonstrates that fibre reinforcement can slow the diffusion of reagents into the polymer matrix; although works describe that the fibre reinforcement can also facilitate greater mass transfer than solid polymers by providing a pathway for diffusion [38].

To avoid the use of alkaline conditions, high-temperature methanolysis of PET can be used to generate DMT as the major component, with a yield of up to 97.7%. Here, it has been suggested that two methanol molecules break two ester linkages at temperatures in the range of 280–310 °C, which results in the formation of carboxymethyl groups [39]. Although not the original monomer, DMT is still widely used as a precursor in the manufacturing of industrial plastics, and so its recovery still represents an improvement in resource efficiency compared to landfilling or incinerating composites [40].

In addition to PET, alcoholysis has also been applied to PLA, most commonly using methanol. Although it has not yet been widely studied in fibre-reinforced composites, the research which does exist may be applied to FRPs, especially with regard to the organic products generated. In the case of methanol, methyl lactate is obtained, with acids, bases, and weak-Lewis acids investigated as potential catalysts. In the latter case, an organic iron chloride salt ([Bmim]FeCl$_4$) enabled the recovery of up to 95% methyl lactate [41].

As with hydrolysis reactions, high temperatures in excess of 250 °C are often needed to solubilise high-performance thermoplastics, regardless of whether or not a catalyst is used. These materials, typically polyamides, PPS resins, sulphone polymers, or aromatic polymers, often have strong aromatic backbones, necessitating harsher conditions to achieve sufficient bond cleavage. Here, alcohols can act not just as solvents but also as reactants, producing a range of products in addition to, or instead of, the original monomers. As an example, both methanol and propanol have been employed under sub- and supercritical conditions in the recycling of PA-based composites. When methanol is used at 330 to 370 °C and pressures up to 39 MPa, reaction times of several hours yield a complex product mixture, including ε-caprolactam, N-methyl caprolactam, methyl 6-hydroxycapronate, and other alkylated derivatives [42, 43]. The relatively low yield of the monomer ε-caprolactam (14%) is attributed to further alkylation reactions. In contrast, propanol delivers a much higher yield of up to 91% at process conditions of 370 °C and 22 MPa, with far fewer side products [43]. The higher activation barrier to alkylation in propanol was proposed as an explanation for this improved selectivity. When considering the recycling of polyamides using alcohols, propanol is therefore more able to achieve a fully circular system. This range of different conditions necessary for the alcoholysis of thermoplastic matrices is provided in Table 2.2.

Table 2.2: Alcoholysis conditions and major reaction products for the depolymerisation of different thermoplastics (adapted from [13]).

Polymer matrix	Solvent/catalyst	Conditions	Key organic products	Reference
PET	Methanol/NaOH or KOH	110–150 °C, >10 min	Terephthalic acid Ethylene glycol Dimethyl terephthalate	[37]
PET	Methanol	280–310 °C, 50–70 min 8–10 MPa	Dimethyl terephthalate	[39]
PLA	Methanol/[Bmim] FeCl$_4$	120 °C, 3 h	Methyl lactate	[41]
PA	Methanol	330–370 °C, 6 h 27–39 MPa	ε-Caprolactam, N-methyl caprolactam, Alkylated derivatives	[42, 43]
PA	Propanol	370 °C, 22 MPa	ε-Caprolactam	[43]

2.2.3 Glycolysis

Like alcoholysis, glycolysis is a transesterification process where diols, most commonly EG, cleave ester linkages in thermoplastic matrices to yield lower-molecular-weight oligomers and, ideally, monomeric species. For fibre-reinforced thermoplastics, glycolysis has been most widely studied for PET, while other commodity and engineering thermoplastics have received less attention. For PET, glycolysis is well established, with EG and a catalyst readily producing bis(hydroxyethyl) terephthalate (BHET) [17]. However, the product mixture is also likely to contain PET oligomers, making the separation of pure BHET difficult, although membrane-reactor systems have been developed [44]. Catalyst choice influences both rate and selectivity; metal acetates, specifically zinc acetate, are amongst the most widely studied. In these systems, temperatures and pressures in the range of 180–250 °C and 0.1–0.6 MPa have been applied, usually over a time period of 30 min to several hours [17]. More recently, advancements in this field have combined the previously described DBU catalyst with *para*-toluene sulphonic acid (*p*-TSA). Although this has yet to be applied to fibre-reinforced composites, this system was able to facilitate the glycolysis of PET with complete conversion in 72 min at 180 °C [45].

In addition to PET, PLA may also undergo glycolysis. In one study, PLA was treated with polyethylene glycol (PEG) at ~220 °C, producing PLA/PEG block copolymers following solvent exchange and precipitation [46]. This route does not regenerate lactic acid monomers but instead yields new polymeric products with potential value. Whereas methanolysis will produce methyl lactate, glycolysis does not, thus highlighting that the choice of reactant/solvent system influences product distribution.

2.2.4 Other reaction systems

In addition to the processes described above, less conventional reaction systems have been developed, often for specific polymer types. For example, a relatively low-temperature system for the recycling of carbon fibre-reinforced polyamide has been reported. In this, the CFRP was soaked in benzyl alcohol at 160 °C for 60 min. Subsequent cooling and the addition of acetone resulted in a polymeric precipitate, which could be reused in a new composite along with the reclaimed fibres. However, mechanical testing of the secondary composite demonstrated a 40% reduction in both tensile strength and stiffness compared to the virgin material. The authors attributed this to the non-aligned fibres agglomerating, which caused areas of localised stress within the composite and hence early failure during testing [47].

Carbon fibre-reinforced polyamides have also been depolymerised through novel solvent systems. Supplied by CreaCycle GmbH, the exact formulations are not available but have been described as non-polar, non-polar aprotic, polar aprotic, or polar protic. Based on their Hansen solubility parameters, only two polar protic solvents were matched to the decomposition of the polyamide. At temperatures in the range of 140–180 °C, it was possible to fully solubilise the polyamide within 0.75 to 3 h. Up to 99.8% of the initial solvent was recovered, along with clean fibres. The recovered polymer showed only a 10% reduction in its average molecular weight when compared with virgin material. There was also only a slight decrease in the single fibre tensile strength reported, thus suggesting that this solvent system has significant potential for the complete chemical recycling of carbon fibre-reinforced polyamides [48].

Polycarbonates have thus far not been significantly discussed in this section, but their solvolysis has been recently investigated [49, 50]. One study demonstrated a sonochemical solvolysis process using task-specific ionic liquids within a 2-methyltetrahydrofuran and methanol solvent system, enabling up to 84% polymer conversion and 81% bisphenol A (BPA) yield at just 30 °C [50]. Complementary work developed an ionic liquid-catalysed transesterification of waste PC using dimethyl carbonate as a sustainable methylating agent and N-methylpyrrolidone as a co-solvent, achieving 100% depolymerisation and 99% selectivity toward bisphenol A dimethyl ether (BPAME) at 150 °C within 6 h [49]. Together, this research highlights the development of novel solvent systems, which may have further applications across other polymer types.

Additional research investigated polypropylene (PP) reinforced with carbon fibres. Here, the PP was dissolved in xylene at 135 °C for 60 min which enabled the recovery of carbon fibres. The liquid obtained was cooled, resulting in the precipitation of a solvated PP powder. The addition of acetone at this stage aids in the precipitation of PP and minimises the amount of xylene present in the precipitate. Subsequent filtration and drying of this precipitate resulted in 90 to 93% of the original PP being recovered. This recycled PP was used as the matrix material in the manufacture of a secondary CFRP, the tensile strength of which was similar to that of a virgin sample. Additional testing of the polymer demonstrated that the degree of crystallinity was also largely unchanged following two recycling loops. However, there was a reduction in the average and modal molecular weights of the polymer between the first and second recycling loops. It was thought that this led to a slight decrease in the shear strength of the CFRP [51]. Earlier work has also demonstrated the efficacy of xylene as a solvent when chemically recycling glass fibre-reinforced polypropylene under the same conditions [52]. Use of the recovered glass fibres alongside the recovered polymer in a recycled composite showed similar tensile strength, modulus, and impact strength when compared to virgin material. Both research articles [51, 52] therefore demonstrated that a closed-loop chemical recycling process of fibre-reinforced PP is possible; however, there may be a limit to the number of recycling loops which are achievable without significantly downgrading the mechanical properties of the material. This means that, in order to maintain high product quality, the introduction of some virgin polymer may be necessary upon each remanufacturing stage.

This addition of virgin material will almost certainly be needed when chemically recycling other polymers. For example, in the recovery of glass fibres from a methacrylate-based thermoplastic, some incomplete precipitation of the polymer was observed, resulting in a measured loss in resin of 4 wt.%. It was assumed that this remained stuck to the fibre surface or within the equipment used. In this particular system, chloroform was used to initially dissolve the resin, which was subsequently precipitated in methanol. The polymer solution was dried, which allowed the recovery of most of the original methacrylate resin which, the authors suggest, is an economically viable process under certain conditions [53]. The recycling of poly(methyl) methacrylate (PMMA)-based resins, such as those under the trademark of Elium®, has also been the subject of some academic research. Solvents such as trichloromethane, trichloroethylene, 1,4-dioxane, cyclohexanone, acetophenone, tetrahydrofuran, and dimethyl formamide have all been investigated in the temperature range of 30–70 °C. Polar solvents, such as acetone and ethyl acetate, are most suited to the dissolution of PMMA-type polymers [54]. Indeed, these two solvents, amongst others, have been applied to the recycling of a carbon fibre-reinforced Elium® 150 resin. Although it was possible to completely dissolve the resin at room temperature, long processing times of 24 h were needed [55].

From this section, it is clear that the chemical recycling of fibre-reinforced thermoplastics encompasses a diverse set of solvolytic strategies (hydrolysis, alcoholysis, glycolysis, and others) that have each demonstrated the potential to liberate fibres and recover valuable organic products. Hydrolysis remains the most mature and widely studied, particularly for PET and polyamides, where both acid- and base-catalysed systems can achieve high monomer yields, such as TPA or ε-caprolactam, under controlled conditions. Alcoholysis offers complementary advantages: low-temperature methanolysis of PET and PLA yields useful precursors such as DMT or methyl lactate, while high-temperature, high-pressure systems enable the depolymerisation of tougher engineering polymers such as PEEK, albeit with product streams comprising predominantly smaller platform chemicals (phenol or phenolic derivatives) rather than pristine monomers. Glycolysis can yield BHET suitable for polymer re-synthesis from PET, although separation from oligomeric by-products remains a barrier to closed-loop recycling. Collectively, these approaches illustrate both the opportunities and challenges in adapting solvolysis to reinforced thermoplastics. Fibre integrity, product purity, and scalability are key considerations, which will be discussed further towards the end of this chapter. The following section instead summarises the available literature on the chemical recycling of fibre-reinforced thermosets.

2.3 Recycling of fibre-reinforced thermosets

As described previously, the key difference between thermosetting resins and thermoplastics is the presence of additional small molecules holding the long polymer chains together. These molecules are referred to as hardeners or cross-linking agents, and they form chemical bonds with the polymer chains in a process known as curing. This results in a material with a three-dimensional structure, which does not soften or melt upon heating, although they can be thermally degraded at high temperatures. It is estimated that, of all FRPs currently in circulation, ~80% use a thermosetting polymer as the matrix material [56, 57]. For this reason, it is likely that the majority of waste FRP recycling will involve the decomposition of a thermoset rather than a thermoplastic.

There is a wide variety of thermosetting polymers in use, although some common types include epoxies, unsaturated polyesters, vinyl esters, phenolics, and polyamides, all of which may be cross-linked with amines, peroxides, or thiols. This represents a significant challenge with regards to the circular economy, as with thermosets, it is unlikely that the same chemical bonds formed during curing are subsequently broken upon recycling the material. As a result, it is extremely unlikely that the original monomers are recoverable, and instead, a complex mixture of different compounds is generated. Partly for this reason, individual constituents are rarely identified in the scientific literature; instead, the focus is on characterising the general class of compounds present and whether they can be incorporated into a new resin. Similar to thermoplastics, however, the products obtained are largely dependent on the original matrix constituents, the solvents or catalysts chosen, and the operating conditions used. Previous works have categorised processes based on these operating conditions [12, 13]; however, this section will instead consider the mechanism taking place, which is typically one of hydrolysis, alcoholysis, aminolysis, or acetolysis, although novel solvent systems have also emerged in recent years.

2.3.1 Hydrolysis

Hydrolysis is one of the most extensively studied approaches for the chemical recycling of fibre-reinforced thermosets (FRTSs). The process involves the cleavage of ester, ether, or amide linkages in thermoset matrices through a reaction with water, often under elevated temperature and pressure. Depending on the reaction medium, hydrolysis can proceed with water alone, organic solvent-water mixtures, acidic catalysts, alkaline catalysts, or in the presence of alternative catalytic systems.

2.3.1.1 Water only

Due to its low cost, availability, and relative lack of hazards, water was one of the first solvents considered for the chemical recycling of FRTSs and has since been applied to a wide range of composites. Alone, sub- or even supercritical conditions are needed. In this phase, it is considered an adjustable solvent which is able to support ionic, polar non-ionic, and free radical reactions depending on the process conditions [58]. Unfortunately, water has a high critical point, beyond that of other GRAS solvents, as shown in Table 2.3, and thus, unless subcritical conditions are sufficient for the recycling process, the energy demand is likely to be high. In addition, supercritical water is corrosive to non-nickel-based alloys, which also leads to expensive process equipment; reactors, fittings, and associated pipework may need to be constructed of specialist material such as Hastelloy stainless steel.

Table 2.3: Critical points of solvents generally recognised as safe (GRAS) (data taken from [59]).

Solvent	Critical temperature, T_C (°C)	Critical pressure, p_c (MPa)
Water	374	22.1
Methanol	240	7.95
Ethanol	241	6.30
Propan-1-ol	264	5.20
Acetone	235	4.80

Whether or not supercritical conditions are necessary is largely dependent on the polymer. For example, temperatures in excess of 400 °C are needed for thermoset epoxies [60, 61]; however, subcritical temperatures of less than 300 °C are sufficient for unsaturated polyesters [58]. A summary of some of the reaction conditions used in the recycling of thermoset FRPs with water is provided in Table 2.4.

Table 2.4: Selected reaction conditions for the degradation of various thermoset FRPs using water only (adapted from [13]).

Polymer type	Temperature (°C)	Pressure (MPa)	Time (min)	Source
Unsaturated polyester	275	6	40	[58]
Amine-cured cresol/BPA epoxy	400 °C	27	30	[61]
RTM6 epoxy	375 °C	25	15	[62]
Epoxy resin cured with phthalic anhydride	440 °C	30	35	[63]
BPA-type epoxy cured with isophorone diamine	290 °C	Not given	75	[64]

It is evident that, under supercritical conditions, reactions proceed quickly, with clean fibres liberated in less than 1 h. As expected, the primary organic products are dependent on the formulation of the polymer matrix but commonly include cyclic compounds such as phenol, benzene, and their derivatives, as well as larger oligomers [58]. A similar approach applied to glass fibre-reinforced unsaturated polyesters showed complete resin decomposition at 450 °C, although significant fibre damage was observed [65]. These results confirm that water alone is capable of matrix breakdown under extreme conditions, but fibre preservation, particularly for glass, is a challenge, as discussed in later sections of this chapter.

2.3.1.2 Organic solvent mixtures

In an effort to reduce the required temperature and pressure (and hence achieve a reduction in energy demand and equipment cost), some research has focused on combining water with either acetone or short-chain alcohols. As shown in Table 2.5, the critical points of these organic solvents are substantially lower than that of water, meaning that the reaction system will benefit from supercritical fluid properties under milder conditions. The optimum solvent-to-water ratio depends on both the solvent and the polymer type. For example, for a model epoxy resin based on BADGE, 80 vol.% acetone and 20 vol.% water achieved maximum decomposition [66]. This solvent system has also been applied to the recycling of real end-of-life waste at a wide range of conditions. An RTM6 epoxy of unknown formulation, reinforced with carbon fibre was degraded at 320 °C in 120 min. This system showed a strong temperature dependency, with an increase to just 340 °C resulting in a reduction in reaction time to 45 min [67].

Alternative solvent systems have demonstrated similar success. In one study, a 50:50 vol.% ethanol–water mixture was used to recycle a comparable FRP. Under semicontinuous flow conditions at 375 °C for 15 min, 96% of the resin was removed [68]. The enhanced mass transfer associated with the flow reactor likely contributed to the reduced processing time and temperature. These examples highlight the strong influence of resin chemistry, solvent choice, and reactor configuration on process efficiency. However, limited data make it difficult to decouple the benefits of semicontinuous operation from solvent and resin effects. While semicontinuous reactors can offer improved performance, they are inherently more complex and costly than batch systems. As such, considerable research has focused on the development of catalysts to reduce time and temperature requirements for the hydrolysis of FRTSs. As explored in the following sections, however, a balance must be sought between the aggressive nature of very strong acids and bases, which deliver clean fibres at less than 200 °C, and these GRAS solvents, which demand more extreme conditions.

2.3.1.3 Acid-catalysed hydrolysis

Acid catalysis has been widely investigated as a strategy to accelerate the hydrolysis of FRTSs. Protonation of ester, ether, or amide bonds within the matrix reduces the activation barrier to bond cleavage, enabling depolymerisation under milder conditions than would be possible with water alone. Both mineral and organic acids have been studied, spanning low-temperature reflux systems through to high-temperature, high-pressure processes.

Strong acids such as hydrochloric acid (HCl), sulphuric acid (H_2SO_4), and nitric acid (HNO_3) have been shown to promote resin degradation at temperatures as low as 80 °C. For example, 4 M HNO_3 solubilised an amine-cured epoxy matrix at this temperature, generating aromatic amines and phenolic species as the major organic products [69]. However, long reaction times of 100 h were needed, which may limit the economic viability of a commercial process. Higher concentrations and slightly higher temperatures have been shown to accelerate the recycling process in other epoxy-based composites. In other examples, 8 M HNO_3 achieved close to 100% degradation in just 5 h at 90 °C [70], while in other research, 12 M HNO_3 required 6 h at the same temperature [71]. The main difference between these two reaction systems is likely the composite loading. In the former example, this stood at 40 g L^{-1}, while in the latter it was increased to 55 g L^{-1}. The higher composite-to-solvent ratio means that there is likely to be greater mass transfer limitations.

Similarly, sulphuric acid may be used under similar conditions to deliver acid-catalysed hydrolysis of FRPs. At 110 °C, an unspecified concentration of H_2SO_4 was used alongside hydrogen peroxide (H_2O_2) to degrade an anhydride-cured epoxy. The system required stirring and still took several hours [72]. Unsurprisingly, increasing the temperature leads to significantly faster reaction rates. Additional research showed that 1 M H_2SO_4 in water is able to decompose a BPA-based epoxy at 260 °C in just 15 min.

However, the authors highlighted that some surface degradation of the fibre reinforcement was observed, which in turn may reduce their mechanical performance [73].

Interestingly, relatively low concentrations of HCl may facilitate fibre recovery. In one report [74], just 0.1 M HCl was needed in a water/tetrahydrofuran system to degrade a cresol-based epoxy resin cured with vinyl ethers at room temperature. Although the reaction time was relatively long at 24 h, this is still significantly quicker than the 100 h described previously with HNO_3. However, it is worth noting that due to differences in polymer formulations, reactor configurations, and loadings, it is not possible to conclusively state which recycling system enables the most efficient fibre recovery. It is also worth noting that these strong acids may have a detrimental effect on the properties of the recycled fibres, as discussed in Section 2.4.

In addition to strong mineral acids, it is apparent that weak organic acids may also catalyse the degradation reaction. A combination of acetic acid with *p*-TSA and water was able to degrade an unsaturated polyester resin at 180 °C. After a relatively long reaction time of 12.5 h, ester bonds throughout the network were successfully broken to produce glass fibres [75].

As these examples show, acid-catalysed hydrolysis offers a route for the chemical recycling of FRTSs. Strong mineral acids are effective depolymerisation agents across a range of conditions, though they tend to produce mixed aromatic streams rather than monomers suitable for direct repolymerisation. Acetic acid, while milder, may offer greener processing alternatives compared to other strong acids; it is safer to handle and can be manufactured from bio-based resources. Fibre quality is also a key consideration: while carbon fibres can often be recovered intact or with beneficial surface functionalisation, glass fibres are more susceptible to attack and may suffer significant strength loss, as described in detail later in this chapter.

2.3.1.4 Base-catalysed hydrolysis

As an alternative to acids, base-catalysed hydrolysis has also been explored for the recycling of FRTSs, although this is less widely researched than acidic media. Further, where acids can be used at relatively mild conditions, high temperatures and pressures are often needed to fully hydrolyse a thermoset matrix when using alkaline catalysts. Similar to H^+ ions, it is the hydroxyl (OH^-) ions which cleave ester and ether linkages within polymer networks, although the precise conditions required are strongly dependent on the polymer type, fibre reinforcement present, and the addition of other reagents. For example, the inclusion of phenol as a co-solvent in conjunction with KOH enabled significant reductions in severity compared to water-only supercritical hydrolysis. Whereas pure water required more than 400 °C to degrade the epoxy resin, a mixture of 1.0 to 1.1 M phenol and 0.18 to 0.36 M KOH achieved 95% resin decomposition at 315 °C in only 30 min [76]. The synergistic effect of phenol was attributed to the generation of phenoxy radicals; however, both insufficient and excessive phenol concentrations inhibited decomposition, either by limiting radical availability or by promoting phenol migration into the epoxy network [76].

It is worth noting that the ability of alkaline media to accelerate the degradation reaction is also dependent on the specific solvent used. In other work exploring an acetone–water mixture for the recycling of a commercial RTM6 epoxy, the addition of both NaOH and KOH at concentrations between 0.01 and 0.4 M had either no effect or hindered the degradation reaction. Further exploration showed that the alkaline salts catalysed the aldol condensation of acetone to form 4-hydroxy-4-methyl-pentan-2-one. Under the conditions explored, at about 300 °C, the original acetone–water mixture was likely to be in the supercritical state. However, the elimination of acetone from the solvent system likely caused an increase in the critical point of the mixture, and hence the reaction no longer benefited from the enhanced properties of SCFs [77]. Together, these works therefore highlight the importance of selecting catalysts which complement and work with the chosen solvent system to avoid the potential for competing reactions.

2.3.1.5 Weak Lewis acids

The above sections have only considered Brønsted–Lowry acids and bases; however, recent research has examined the application of weak Lewis acids in recycling thermoset composites. One strategy for their chemical recycling involves the selective cleavage of the amine (C–N) bonds frequently present throughout the three-dimensional network. Catalysts which can achieve this may be large metal complexes or simple transition metal salts. The former type is often too large to effectively diffuse into FRTSs, but smaller molecules have no such limitations. One such example is $ZnCl_2$, which is also highly soluble in water and alcohols and thus easily disperses throughout a solvent. Under relatively mild temperatures, it is subsequently able to diffuse into an FRP. Here, the Zn^{2+} ion co-ordinates with a nitrogen atom and selectively cleaves the associated amine bond [78]. For example, an aqueous solution of $ZnCl_2$, heated to 210 °C was able to decompose 95% of an epoxy resin and thereby recover clean carbon fibres. However, a high concentration and long reaction time of 11 M and 9 h, respectively, were needed [78]. To reduce the required concentration and reaction time, higher temperatures and co-solvents are necessary. In other research, a temperature of 250 °C, $ZnCl_2$ concentration of 4.4 M, and a reaction time of 5 h was sufficient to recover clean carbon fibres [79]. Here, the principal degradation products were phenolic compounds, consistent with the cleavage of the bisphenol A-derived epoxy backbone.

Further reductions in catalyst concentration have been achieved by Keith et al., who applied a range of metal chloride salts to the degradation of a commercial thermoset CFRP using an acetone-water system [77]. Solutions of 0.05 M $ZnCl_2$, 0.05 M $MgCl_2$, and 0.005 M $AlCl_3$ facilitated complete resin removal and the recovery of carbon fibres. By optimising time and temperature, complete degradation was achieved at 290 °C in 90 min or 300 °C in 45 min, representing a 40 °C reduction compared with the neat acetone–water system. Infrared spectroscopy suggested that the metal ions facilitated cleavage of C=N bonds in the epoxy network, while gas chromatography identified cyclic degradation products and low concentrations of amine derivatives [77]. Together, these studies highlight the potential of weak Lewis acids to play a role in the hydrolytic recycling of thermoset composites.

It is not only metal chlorides which act as weak Lewis acids for the degradation of polymer composites. Zinc acetate ($Zn(OAc)_2$) has been supplied at concentrations of up to 20 wt.% in water for the recycling of amine-cured epoxies from wind turbine waste. At 250 °C, 83.5% degradation was achieved in 5 h. Although this is a relatively low rate of degradation, the authors demonstrated that the recovered glass fibres and decomposed matrix could be directly melt-compounded with a polyamide at up to 70 wt.% to give a new fibre-reinforced thermoplastic with a high tensile strength and modulus of 131.3 MPa and 15.5 GPa, respectively [80]. The $Zn(OAc)_2$ catalyst was also recovered, and this work demonstrates the possibility of a hydrolysis-based closed-loop recycling process under relatively mild conditions.

2.3.2 Alcoholysis

Alcohols, typically methanol, ethanol, and propanol, have widely been applied to the chemical recycling of fibre-reinforced thermosets. Compared with hydrolysis, alcoholysis offers the advantage of using solvents that achieve supercritical conditions at lower temperatures and pressures, thus reducing energy requirements and reactor costs. They are also generally recognised as safe (GRAS) and can be manufactured from biobased, rather than fossil-derived, sources. These alcohols have typically only been investigated at the lab-scale, either neat or in combination with various catalysts. The process typically targets the cleavage of ether and amine linkages within the polymer network, producing a range of oligomeric and aromatic compounds. While fibre liberation is usually successful, differences in resin, solvent properties, and catalysts strongly influence the required process conditions, degradation rates, fibre quality, and the composition of the organic product stream. An overview of reported experimental conditions and corresponding resin degradation is provided in Table 2.5. Although short-chain alcohols are most commonly investigated, one study applied benzyl alcohol in conjunction with NaOH at 2.6 M. This achieved more than 90% resin removal at 195 °C within 40 min [81].

Methanol is one of the most widely trialled alcohols for the decomposition of FRTSs, owing to its low critical point (240 °C, 7.95 MPa [59]). However, in practice, high temperatures are often required to achieve substantial decomposition, with a strong dependence on the polymer chemistry. In some studies, a carbon fibre-reinforced amine-cured cresol/bisphenol-A epoxy (LTM26EL) only demonstrated a 60% degradation when processed at 450 °C, 10.6 MPa for 15 min [82]. In contrast, a bisphenol-A epoxy cured with cyclic anhydrides achieved near-complete degradation at considerable milder conditions of 350 °C in the same time of 15 min [83]. This highlights the importance of resin formulations, as variations in cross-linking density and curing agents directly impact alcohol penetration and bond cleavage. Furthermore, it exemplifies the difficulty in comparing the efficacy of different solvent systems when there is a wide range of thermoset matrices described in the literature.

Similar to methanol, the relatively low critical point of ethanol (241 °C, 6.3 MPa [59]) makes it an attractive solvent. Again, successful depolymerisation depends on resin chemistry and the use of catalysts. In the same study using LTM26EL epoxy [82], ethanol at 450 °C enabled a degradation of 79% in 15 min, compared to the 60% achieved using methanol. Although this is higher, it is still likely to be insufficient to recover usable fibres; degradations of above 90–95% are often required for fibres to be considered "clean" and suitable for reuse in a new composite material. This was achieved for an amine-cured epoxy by Okajima et al. in a slightly longer reaction time of 25 min. Although here a lower temperature of 320 °C was used, it was worth noting that a higher pressure of 11.2 MPa was induced by the higher solvent loading within the batch reactor [84]. In turn, this higher pressure may result in greater penetration of ethanol into the composite material and hence facilitate a faster degradation rate. In both these studies, quite severe conditions are needed, which in turn is likely to result in expensive process equipment. As such, catalyst-assisted ethanolysis has been explored and does offer a significant improvement [79]. Here, amine-cured epoxy from aerospace waste was degraded using a 1.4 M $ZnCl_2$ catalyst in ethanol. A resin decomposition of ~90% was achieved in 5 h at just 220 °C, offering a 30 °C reduction compared to using $ZnCl_2$ in water. Infrared (FTIR) and nuclear magnetic resonance (NMR) analyses revealed that the decomposition products were primarily oligomers with average molecular weights of ~650 Da, consisting of benzene rings and cyclic structures with methyl and amine functional groups. The polydispersity index of 1.5 indicated a relatively narrow distribution of oligomer sizes. At higher temperatures (220–250 °C, 9 h), smaller molecules and more complex mixtures were obtained, reducing suitability for reuse in new resins. Nevertheless, the oligomeric products generated at lower temperatures were successfully incorporated into new CFRPs at up to 15 wt.% loading, demonstrating the potential for partial closed-loop recycling [79].

In addition to the above examples, propan-1-ol has also been extensively studied for the recycling of FRTSs, with both neat and catalysed systems reported. For the LTM26EL epoxy, 95% of the resin was removed in 40 min using 450 °C and 25.4 MPa [82]. Similarly, complete degradation was achieved in just 25 min at 320 °C and 9 MPa for another amine-cured epoxy [84]. Due to their simplicity, the equipment used in these works was relatively simple batch reactors. Occasionally, however, semi-continuous flow reactor systems have been reported [68, 85]. Here, a composite sample is loaded into the reactor, and the solvent is pre-heated, pressurised, and pumped into the reactor before being continuously removed. By supplying fresh solvent, a greater concentration gradient is maintained than in batch systems, which may help overcome mass transfer limitations. In one study, an epoxy resin pre-preg material was processed with propanol in such a system at 310 °C and 5.2 MPa for a total time of 40 min. This enabled the recovery of clean fibres (as determined by scanning electron microscopy) with similar mechanical properties to the as-received material [85]. Combining these semi-continuous flow reactor systems with alkaline catalysts further accelerate the degradation process. The strong alkalis KOH, NaOH, and CsOH were combined with propanol and used to successfully recycle carbon fibres at a low concentration of just 0.06 M in just 15 min at 300 °C under batch conditions [82]. Here, the OH^- ions form free radicals and also promote the dehydrogenation of propanol. This provides an additional mechanism for resin degradation through subsequent hydrogenation of the polymer, which may also lead to reduced char formation [86]. In a semi-continuous flow reactor, even lower concentrations of 0.02 M KOH and lower temperatures of 275 °C were sufficient to degrade the resin and recover usable fibres. Although a longer reaction time of 70 min was needed, this research highlights how combining different approaches can deliver significant reductions in process conditions. In turn, this may improve both the economic viability and environmental impact of a commercial process.

Table 2.5: Reported conditions for the alcoholysis of different FRTSs and the corresponding achieved resin degradation (adapted from [13]).

Polymer type	Solvent/ catalyst system	Temperature (°C)	Pressure (MPa)	Time (min)	Degradation	Source
LTM26EL epoxy	Methanol	450	10.6	15	60%	[82]
Anhydride-cured BPA epoxy	Methanol	350	~9.0	15	~100%	[83]
LTM26EL epoxy	Ethanol	450	8.0	15	79%	[82]
Tetra-ethylene pentaamine-cured BPA epoxy	Ethanol	320	11.2	25	95%	[84]
Amine-cured epoxy	1.4 M $ZnCl_2$ in ethanol	220	Not given	300	~90%	[79]
LTM26EL epoxy	Propanol	450	25.4	40	95%	[82]
Tetra-ethylene pentaamine-cured BPA epoxy	Propanol	320	9.0	25	~100%	[84]
Cured epoxy	Propanol	310	5.2	40	Clean	[85]
LTM26EL epoxy	0.06 M KOH in propanol	300	8.9	15	86%	[82]
LTM26EL epoxy	0.02 M KOH in propanol	275	Not given	70	97%	[82]
Anhydride-cured BPA epoxy	2.6 M NaOH in benzyl alcohol	195	Atmospheric	60	~95	[81]

2.3.3 Aminolysis

Aminolysis is an emerging pathway for the chemical recycling of FRTSs. It involves the use of amines (either primary, secondary, or polyfunctional molecules) as solvents and reactants to cleave the cross-linked polymer network. The process typically targets ester and amide bonds, but amines can also attack ether linkages or coordinate with nitrogen atoms in an epoxy backbone, thereby promoting more selective bond scission than water and alcohols. Aminolysis may also have an additional advantage over hydrolysis and alcoholysis through the generation of nitrogen-containing degradation products, such as amides and amines, which may serve as functional intermediates for new resin formulations. Research has primarily focused on epoxy matrices, although unsaturated polyesters have also been investigated, as described in this section.

Monoethanolamine (MEA) has been identified as both a solvent and a reactant in the recycling of FRTSs. In one example, an anhydride-cured epoxy was completely degraded at 160 °C within 60 min using MEA in conjunction with 0.5 M KOH [87]. The combination facilitated rapid solubilisation of the resin, producing a mixture of complex amides and hydroxy-functional compounds as a result of the reaction between the epoxy matrix and MEA. The products were sufficiently reactive to be incorporated into a new resin system without extensive downstream processing, although mechanical testing of the secondary composite showed mixed results: flexural modulus increased, while flexural strength decreased by ~50%, attributed to incomplete curing at the relatively low curing temperature used. Fibres were reported to be cleanly liberated with minimal surface damage, suggesting MEA is an effective medium for fibre recovery under mild conditions [87].

Polyamines, such as diethylenetriamine (DETA), have also been explored as recycling agents. DETA acts as both a strong nucleophile and a swelling agent, promoting penetration of the resin network and efficient bond cleavage. In one study, an anhydride-cured epoxy was pre-treated with dichloromethane (DCM) to generate pores within the matrix [88]. This was then exposed to microwave heating at 130 °C in the presence of DETA to achieve near-complete degradation within 50 min. The reaction mechanism was attributed to the amination of ester bonds by DETA, producing amides and other nitrogen-rich compounds. Importantly, the recovered organics were directly reused in a new resin system, which demonstrated good mechanical properties, underlining the value of amine-derived degradation products. However, it is worth noting that these were only incorporated at up to relatively low proportions of 30 wt.%; beyond this and the thermal properties of the secondary composite were significantly impacted [88]. Fibre recovery was, however, successful, with SEM imaging showing smooth surfaces comparable to virgin carbon fibres. However, the requirement for DCM pre-treatment may limit future scalability due to handling and environmental concerns.

2.3.4 Glycolysis

Glycolysis is a widely researched chemical recycling strategy where hydroxyl-containing glycols penetrate and cleave polymer networks, breaking ester or ether bonds in the resin. It is more commonly associated with the recycling of PET [45, 89], but it has also been applied to thermoset epoxy-based matrices [90, 91]. The advantage of glycolysis lies in its ability to depolymerise cross-linked structures under less extreme conditions than hydrothermal methods, while still yielding recoverable fibres and potentially reusable liquid products. PEG is particularly attractive as a glycolysis agent due to its high boiling point, ability to solubilise polymer chains, and reactivity with both ester and ether bonds.

One of the earliest studies to demonstrate the glycolysis of epoxies with PEG was carried out by Yang et al. [92]. An anhydride-cured epoxy system based on diglycidyl ether of bisphenol A (DGEBA) was investigated. By supplying PEG (Mr = 200) in conjunction with 0.5 M NaOH, the resin was dissolved within just 50 min at 180 °C. This relatively mild condition is notable when compared to hydrolysis or alcoholysis routes, which often require higher temperatures or harsher solvents. The combination of PEG and NaOH acted synergistically: the PEG molecules penetrated the resin network, while the alkaline medium promoted the cleavage of ester bonds, yielding oligomeric products containing hydroxyl functional groups. Other matrix materials may, however, demand harsher conditions. In other work, complete solubilisation of an epoxy resin required a reaction time of 4 h at 200 °C, using a significantly higher concentration of NaOH (5.6 M) [90]. This indicates that the presence of fibre reinforcement hinders penetration and slows resin decomposition. Similarly, macrogol 400, a lower molecular weight PEG, has been trialled with KOH as a catalyst [91]. Here, CFRPs were pre-treated with nitric acid and ultrasonication in acetone before immersion in the macrogol/KOH solution at 160 °C for 200 min, leading to near-complete resin removal. Although effective, the requirement for multi-step pre-treatment limits scalability. These works again highlight the variability between resin formulations, as different curing agents or cross-link densities will influence degradation rates. The molecular weight of the PEG should also be considered. Despite the longer reaction time, clean fibres were recovered in all cases, suggesting that PEG-based glycolysis remains a possible route for epoxy recycling.

2.3.5 Acetolysis

Acetic acid has been investigated as both a solvent and a reactive medium for the decomposition of fibre-reinforced thermosets. Its use is attractive for several reasons: it is widely available, may be produced from bio-based resources, is relatively inexpensive, and offers the ability to swell polymer matrices and hence facilitate solvolysis under both low- and high-temperature conditions. Alone, it predominantly acts as a swelling agent, and to achieve sufficient resin degradation, it must be combined with oxidising agents, metal chlorides, or other solvents.

The most commonly investigated oxidising agent is hydrogen peroxide (H_2O_2). In one example, an acetic acid pretreatment involving reflux at 120 °C for 30 min, followed by immersion in dimethyl formamide (DMF) and 30 vol.% H_2O_2 at 90 °C, degraded an epoxy resin and liberated high-quality carbon fibres [93]. However, extensive washing and drying over a 12.5 h-period was needed to fully remove the organic material. A similar approach using acetone/H_2O_2 showed comparable efficiency, again relying on acetic acid as a pre-treatment solvent to open the resin network [94]; however, it should be noted that these mixtures are extremely hazardous. In another study, a mixture of acetic acid and H_2O_2 alone was sufficient to recycle an aerospace-grade CFRP at 65 °C within 4–5 h. Here, the ratio of acetic acid to H_2O_2 was varied, and a composite-to-reactant loading of 17 g L^{-1} was employed [95]. These works highlight the swelling capacity of acetic acid and the oxidative cleavage enabled by H_2O_2 as a powerful combination for FRTS recycling under relatively mild conditions.

To avoid strong oxidising agents, metal chlorides have again shown some promise. $AlCl_3$, combined with acetic acid at a concentration of 1.4 M, enabled the recovery of clean carbon fibres from an epoxy matrix in 6 h at 180 °C [96]. The mechanism here differs from oxidative acetolysis; acetic acid enhances resin swelling and facilitates the diffusion of Al^{3+} ions into the matrix, where they selectively cleave C–N bonds. This route is particularly promising for epoxy systems, which rely heavily on amine curing. It is, however, worth noting that elemental analysis showed that 13 wt. % of the products contained chlorine, suggesting that chlorocarbons are created due to the presence of $AlCl_3$. This organic mixture may therefore need to undergo dechlorination steps before the decomposition products can be reused.

Acetic acid has also been trialled for the recycling of unsaturated polyester (UP) composites. Zhang et al. employed a mixed system of *p*-toluene sulphonic acid (*p*-TSA) with acetic acid and water at 180 °C for 12.5 h [75]. This resulted in the cleavage of the ester bonds, consistent with an acetolysis mechanism, yielding styrene–maleic anhydride copolymer, phthalic acid, and ethylene glycol diacetate. Although the yields of some products were relatively low due to acetic acid retention, the styrene–maleic anhydride fraction demonstrated potential as an emulsifier for use in other chemical industries.

Although acetic acid may be considered a green solvent, there is increasing attention on exploring further alternative, safe, non-toxic methods for the solvolysis of FRTSs. This includes the application of deep eutectic solvents (DESs), as described in the following section.

2.3.6 Deep eutectic solvents

DESs are a form of ionic liquid which, when combined, form a eutectic mixture with a melting point far lower than that of either individual component. They consist of a hydrogen bond acceptor (HBA) and a hydrogen bond donor (HBD) and are considered a "green" alternative to the wider category of ionic liquids because they may be bio-derived and biodegradable. Furthermore, they are inexpensive, tuneable in polarity and hydrophobicity, and therefore able to support a wide range of reactions and separations [97]. The recycling of FRPs is, at the most fundamental level, both a separation and a reaction; the fibre reinforcement must be separated from the polymer matrix, and to achieve this, a reaction which involves breaking polymer bonds is needed.

One of the most effective systems studied to date is a hydrophobic DES consisting of thymol (HBA) and decanoic acid (HBD) in a 1:1 molar ratio, combined with zinc chloride ($ZnCl_2$) as a catalyst [98]. This formulation required only 3.3 wt.% $ZnCl_2$, far less than the 20–60 wt.% typically used in aqueous or ethanol-based solvolysis systems. The reaction was conducted at 180 °C under atmospheric pressure with a total reaction time of 1.5 to 2 h. Under these conditions, the CFRP epoxy matrix was fully degraded, representing a significant reduction in energy demand compared to conventional high-temperature and pressure processes.

The decomposition mechanism was attributed to the selective cleavage of C–N and C–O–C bonds in the epoxy resin by Zn^{2+} ions, facilitated by the DES medium. Characterisation of the liquid organic fraction by MALDI-TOF/MS and FTIR confirmed the formation of bisphenol A and a range of phenolic derivatives, with molecular weights below 700 Da. These lower-molecular-weight oligomers and monomers contrast with the higher molecular weight degradation products (1,200–2,600 Da) commonly observed in supercritical methanolysis [83], demonstrating the improved depolymerisation efficiency of DES systems. The conservation of aromatic structures suggests potential value recovery from the organic fraction, either for incorporation into new resins or as platform chemicals. A further advantage of DES-based recycling is the reusability of both the solvent and catalyst. In this system, $ZnCl_2$ and the thymol–decanoic acid mixture could be regenerated and reused several times without a significant loss in decomposition efficiency. This recyclability of the solvent system, combined with mild operating conditions, positions DESs as a promising platform technology for the recycling of fibre-reinforced thermosets.

The previous sections have summarised a range of different solvents, catalysts, reaction systems, and conditions which are able to break down polymer matrices and, hence, recycling fibres and potentially the resin constituents. However, most of this work has been conducted on a lab scale, and for the design of a commercial-scale process, it is essential to understand the kinetics of the degradation reaction. To this end, numerous works have conducted kinetic studies, which are outlined below.

2.3.7 Reaction kinetics

For any chemical reaction, the speed, or rate, at which the reaction occurs is governed by the reaction kinetics. Understanding the kinetics allows the relationship between process time, temperature, reaction order, and concentration to be determined, and thus facilitates the optimisation of a process. There have been a number of studies investigating the kinetics of chemically recycling FRPs. Two main approaches have been considered: either following a conventional rate equation or a shrinking core model (SCM).

A conventional rate equation typically has the form shown in eq. (2.1) This simply states that the rate of reaction ($-r_A$, mol L^{-1} s^{-1}) is proportional to the concentration of one or more reactants ($[A]$, mol L^{-1}) raised to the power of the order of reaction, n (dimensionless). In this case, the constant of proportionality is also referred to as the rate constant, k, the units of which depend on the value of n. For example, when $n = 1$, the units of k are s^{-1}, when $n = 2$, the units of k are L mol^{-1} s^{-1}:

$$-r_A = \frac{d[A]}{dt} = k\,[A]^n. \tag{2.1}$$

Unfortunately, the rate constant is not actually a constant but varies with temperature. This relationship may be expressed by the Arrhenius equation shown in eq. (2.2), or the linear form as given by eq. (2.3). Here, k_0 is the frequency factor (s^{-1}), E_A = activation energy (kJ mol^{-1}), R = universal gas constant (kJ mol^{-1} °C^{-1}), and T = temperature (°C):

$$k(T) = k_0\,e^{-\frac{E_A}{RT}}, \tag{2.2}$$

$$\ln(k) = \ln(k_0) - \frac{E_A}{RT}. \tag{2.3}$$

To calculate values for k and n, it is possible to study the evolution or consumption of one or more reactants. For example, the kinetics of the decomposition of two polyester resins have considered the multiple different reactions taking place. This included modelling dehydration, hydrolysis, and decarboxylation reactions individually to calculate specific values [58]. In addition to its complexity, this approach also relies on knowing the formulation of the resin, which is not always possible, especially when recycling commercially available materials.

Alternatively, the rate of decomposition of the resin can be described by considering the concentration of all degradation products in the solution or the mass fraction of resin remaining as a solid within the composite. The order of reaction may be determined graphically through the integration of eq. (2.1). Repeating the kinetic analysis at different temperatures enables the Arrhenius parameters, k_0 and E_A, to be determined. Detailed methods and Arrhenius plots are available in the literature [61, 67]. A summary of available values for k_0 and E_A is provided in Table 2.6 This shows that there is a wide range of possible values, which are likely due to differences in the resin, solvent, additives, reaction conditions, and assumed reaction order.

In addition to a conventional rate equation, a shrinking core model has also been applied to the chemical recycling of FRPs. This model is fully described by Levenspiel [99]; however, in brief, it consists of the following steps:

Table 2.6: Arrhenius parameters calculated for the chemical recycling of FRPs using a conventional rate equation.

Polymer	Solvent	Catalyst	Reaction order, n	Frequency factor, k_0	Activation energy, E_A (kJ mol^{-1})	Source
Amine-cured cresol/BPA epoxy	Water	None	2	2.44×10^4 g_{resin} g_{water}^{-1} min^{-1}	35.5	[61]
Unsaturated polyester	Water	None	3	48.8 to 105 L^3 mol^{-3} s^{-1}	53 to 56	[58]
LTM26EL-cured epoxy	Propan-1-ol	None	1.5	1.25×10^6 min$^{0.5}$ $g_{propanol}^{0.5}$ $g_{resin}^{-0.5}$	95.6	[82]
RTM6-cured epoxy	Acetone/water (80:20 v/v)	None	1	7.64×10^{17} min^{-1}	222.3	[67]
RTM6-cured epoxy	Acetone/water (80:20 v/v)	0.05 M ZnCl$_2$	1	1.34×10^{13} min^{-1}	157	[77]
RTM6-cured epoxy	Acetone/water (80:20 v/v)	0.05 M MgCl$_2$	1	3.67×10^{21} min^{-1}	250	[77]
RTM6-cured epoxy	Acetone/water (80:20 v/v)	0.005 M AlCl$_3$	1	2.02×10^{15} min^{-1}	180	[77]

1. A reactant molecule moves through the bulk fluid.
2. The reactant molecule diffuses through a film surrounding a solid particle, which, in this case, is the surface of the resin.
3. A chemical reaction then takes place on the surface, leaving behind an "ash" layer of decomposition products.
4. Another reactant molecule then diffuses through this "ash" layer after having diffused through the bulk fluid and the film surrounding the particle.
5. As the "ash" layer accumulates, these products migrate away from the unreacted core, thus reducing the size of the particle and causing it to shrink until it disappears.

The rate equation is then developed from a mass balance by considering one of these steps as rate-limiting. With regards to FRPs, the equation for a reaction rate-limited, spherical particle has been demonstrated to give the closest fit to experimental data. This equation is provided in eq. (2.4), and a summary of the Arrhenius parameters calculated with this method is given in Table 2.7

$$1 - (1 - X)^{\frac{1}{3}} = k_{SCM}t. \tag{2.4}$$

Table 2.7: Arrhenius parameters calculated for the chemical recycling of FRPs using an SCM.

Polymer	Solvent	Catalyst	Frequency factor, k_0 (min^{-1})	Activation energy, E_A (kJ mol^{-1})	Source
Amine-cured BPA epoxy	Acetone	None	3.13×10^9	118	[100]
RTM6-cured epoxy	Acetone/water (80:20 v/v)	None	6.69×10^{13}	183	[67]
RTM6-cured epoxy	Acetone/water (80:20 v/v)	0.05 M ZnCl$_2$	7.87×10^{11}	151	[77]
RTM6-cured epoxy	Acetone/water (80:20 v/v)	0.05 M MgCl$_2$	7.31×10^{18}	225	[77]
RTM6-cured epoxy	Acetone/water (80:20 v/v)	0.005 M AlCl$_3$	2.02×10^{15}	175	[77]

The chemical recycling of fibre-reinforced thermosets has advanced through a variety of approaches, including hydrolysis, alcoholysis, aminolysis, acetolysis, and, more recently, deep eutectic solvent systems. These processes differ in solvent type, catalysts, and operating conditions, but all aim to break down matrices into lower molecular weight compounds to liberate usable fibres. Progress has been made in lowering reaction temperatures, shortening processing times, and improving product recovery, with particular promise shown by metal salt catalysis and DES media. However, the extent to which fibres withstand these processes is critical, and the next section, therefore, focuses on describing the characteristics of the fibrous product.

2.4 Fibre properties

The quality of fibres recovered from chemical recycling processes is critical in determining their suitability for reuse in secondary applications. As such, their properties have also been extensively reviewed in previous publications [13]. While advances in solvolysis, hydrolysis, alcoholysis, and other reaction systems have demonstrated the potential for efficient resin removal, the accompanying impact on fibre strength, stiffness, surface chemistry, and sizing often dictates whether recycled fibres can be reintegrated into high-value composites. Carbon fibres typically exhibit greater tolerance to aggressive chemical and thermal environments than glass fibres, which are more prone to corrosion and leaching of oxides [101]. Nevertheless, conflicting reports exist, reflecting variations in fibre type, resin formulation, and process conditions. The following subsections, therefore, examine the mechanical (strength and stiffness) and surface properties (interfacial shear strength and chemical composition) of recycled fibres, focusing first on carbon fibres and then on glass fibres.

In addition to mechanical and surface properties, the length of recycled fibres strongly influences their reuse potential. Long, continuous fibres are generally preferred for high-performance applications since they provide superior load transfer and reinforcement efficiency in new composites. However, due to the size limitations of reactors, it is often necessary to reduce the size of composites prior to processing. Conventional techniques, such as shredding or hammer milling, produce a broad distribution of fibre lengths, which inherently limits their applicability to non-woven mats, fillers, or low-value products. To avoid such downgrading, emerging approaches, including the HiPerDif method [15, 102] and additive manufacturing techniques [103], have been investigated to realign discontinuous fibres into controlled yarns or preforms. If scalable, these methods may allow short fibres obtained from chemical recycling to retain more of the value associated with continuous fibres, provided that their intrinsic quality is preserved during processing. For this reason, this section provides an overview of the performance seen in chemically recycled carbon and glass fibres.

2.4.1 Carbon fibres

Most carbon fibres available on the market are based on poly(acrylonitrile) (PAN). This is carbonised under different conditions, which subsequently determine the properties of the final carbon fibre product. As such, different grades of carbon fibre exist, and so it would be pertinent to classify recycled carbon fibre according to this grade and its properties post-recycling to maximise its use in secondary applications. Doing so could improve confidence amongst manufacturers, especially for non-safety-critical components such as sports equipment, interior car or aircraft parts, or even laptops [4]. During recycling processes, the reaction conditions, the solvent chosen, and the application of acids, bases, and oxidising agents can all have positive or negative effects on the quality of the fibre recovered. As each reaction system is often unique, with different polymer matrices and different fibre types, it is largely not possible to decouple all these different effects from each other; however, some correlations are apparent, which are summarised below.

2.4.1.1 Thermal and solvent effects

Temperature is one of the most critical variables in determining the properties of carbon fibres recovered from chemical recycling processes. It directly influences the extent of resin degradation, the exposure of fibres to aggressive reaction environments, and the preservation or deterioration of fibre tensile strength and modulus. However, due to the variety of different fibre types, solvents, temperatures, and times used in the published literature, it is not possible to decouple the effects of each of these different parameters on the quality of the recycled material.

The interaction between reaction temperature and exposure time is particularly important, though; short exposures to high temperatures may be less damaging than prolonged treatments at lower temperatures, even when complete resin removal is achieved. This is particularly evident in hydrolytic systems, for example, in subcritical and supercritical water systems, where reaction temperatures between 280 °C and 500 °C have been tested. When fibres were exposed to 280–350 °C for 30 min, reductions in tensile strength of 7–18% were reported, with Raman spectroscopy suggesting that the recycled fibres possessed a less pure graphitic structure than their virgin counterparts [20]. Interestingly, the highest retained fibre strength was sometimes observed at the upper end of the temperature range, for example, at 400 °C, if exposure times were halved. This suggests that higher temperatures may accelerate resin removal and hence limit the time for which fibres are exposed to damaging conditions.

Comparable results have been observed in high-temperature alcoholysis. For example, in propan-1-ol, fibres recovered at 320–360 °C over timescales of 30–180 min showed tensile strength losses of 5–15%. The magnitude of the loss correlated with both temperature and time, with increasing severity as either parameter was raised [104]. Similarly, in butanol systems at 330 °C for 60 min, fibres exhibited a 5% reduction in tensile strength and a 7% reduction in modulus, suggesting that higher alcohol boiling points and longer exposures increase the risk of fibre deterioration [105]. Other alcohols may have an even greater detrimental effect on fibre quality. A system using benzyl alcohol and acetone at 160 °C resulted in a reported loss in tensile strength and modulus of 39% and 25%, respectively, despite a relatively short process time of 60 min [106]. However, in another study involving benzyl alcohol and K_3PO_4, the carbon fibres recycled exhibited a strength above 90% of the original value despite the same process time and a higher temperature of 195 °C [81]. It is not thought that this difference is attributed to the presence of acetone in the former system, as other works using acetone at higher temperatures in excess of 300 °C have also demonstrated the recovery of fibres with similar properties to their virgin counterparts [11]. Similarly, other works have also shown that not all high temperatures are damaging to the fibres. In an ethanol/water system, recycled carbon fibres were subjected to 350–400 °C and displayed a surprising increase in tensile strength of up to 19% compared to virgin fibres at 375 °C [68]. Thermogravimetric analysis (TGA) and scanning electron microscopy (SEM) confirmed complete resin removal, ruling out the possibility of resin residue protecting fibre surfaces. Instead, the improvement was attributed to the removal of weaker graphitic planes or defects from the fibre surface, which otherwise act as stress concentrators [68]. This result suggests that, under controlled conditions, high temperatures may improve fibre properties; however, this should be treated with caution. It may be that these effects are dependent on the fibre type and, when handling mixed composite waste from unknown feedstock, it is likely not possible to predict what the original carbon fibres were. These various and sometimes contradicting results reinforce that solvent choice and elevated temperatures must be carefully matched with process durations and, where possible, fibre type to preserve their quality.

Temperature also affects fibre surface chemistry. At elevated conditions, the concentration of surface oxygen (due to the layer of sizing) tends to decrease, as high pressures and temperatures strip functional groups from the fibre surface [68, 105]. By reducing the abundance of polar groups available for bonding to new resins, there may be a reduction in surface functionality, which has implications for secondary composite manufacture. Some systems, for example, with ethanol or $ZnCl_2$, have been shown to increase surface oxygen concentration, with carboxyl, hydroxyl, and carbonyl groups all being detected [107]. This can enhance interfacial adhesion (as determined by interfacial shear strength tests), and hence, the choice of reaction system must be selected with consideration of the impact on these surface characteristics, in addition to mechanical properties.

As illustrated by these examples, temperature and solvent exert a complex, dual influence on carbon fibre properties during chemical recycling. Moderate increases can accelerate resin removal while maintaining acceptable mechanical properties, but prolonged exposure at high temperatures risks significant losses in tensile strength and modulus. The balance between resin degradation, mechanical integrity, and surface chemistry remains critical, and temperature optimisation must be considered in tandem with solvent selection and catalyst choice.

2.4.1.2 Effect of acids

The use of acids in the chemical recycling of CFRPs has been widely investigated due to their ability to break down cross-linked matrices efficiently. However, their influence on carbon fibre properties is mixed, with outcomes highly dependent on acid type, concentration, and reaction conditions. Strong mineral acids, such as nitric, sulphuric, and hydrochloric acid, are particularly effective at resin removal, but they can also impact fibre integrity. Nitric acid has been shown to recover clean fibres with only minor strength reductions ($\approx 4.4\%$) and slight increases in surface oxygen functionalities, such as carboxyl and carbonyl groups [91]. These changes may improve wettability and interfacial bonding in new composites. By contrast, sulphuric acid is harsher; fibres treated with concentrations between 11 and 18 M at room temperature displayed tensile strength reductions of up to 5.8% and modulus losses of around 5% [108]. Hydrochloric acid, applied at 0.1 M in an acetone–water mixture, yielded fibres with minimal property losses ($\approx 5\%$ reduction in strength and modulus) while maintaining comparable interlaminar shear strength in remanufactured composites [109]. However, it is worth noting here that there is a significant difference in acid concentration; it is perhaps not surprising that an increase in H^+ ions by a factor of at least 200 significantly impacts fibre quality.

Weaker acids, such as acetic and tartaric, often coupled with oxidising agents like H_2O_2, have produced more benign outcomes. In acetic acid/H_2O_2 systems, fibres recovered under controlled conditions displayed strengths similar to virgin material, though excessive oxidant concentrations led to performance reductions [95]. Similarly, acetic acid combined with $AlCl_3$ achieved resin removal with minimal fibre damage ($\approx$2% reductions in tensile properties) while enhancing surface oxygen content [110]. As these examples show, mild systems incorporating weak acids or dilute mineral acids show promise for balancing resin degradation with fibre integrity, while concentrated strong acids risk excessive etching or compromising fibre strength.

2.4.1.3 Effect of bases

The most common bases used to accelerate the decomposition of CFRPs are NaOH and KOH. They have been applied across a range of solvent systems, from water to alcohols and glycols, with temperatures spanning from the relatively mild (e.g., 160 °C [91]) to the much more severe 360 °C [104]. Similar to what has already been described; the effects of these bases on fibre properties depend strongly on the solvent choice, the time and temperature used for matrix degradation, and their concentration. Two example systems which appear to have little effect on the fibre properties involved either a 1.3 M NaOH in water solution [111] or an unspecified concentration of KOH in PEG 400 [91]. In the latter case, clean fibres were recovered in 200 min after 160 °C, which exhibited only a 4.4% and 3.1% reduction in tensile strength and modulus, respectively. Similarly, the authors observed improved wettability, which they attributed to an increase in the surface carbonyl and carboxyl functional groups [91]. Where 1.3 M NaOH was applied, slightly harsher conditions of 180 °C and 8 h were necessary, but this only resulted in a measured 2.4% reduction in strength and no significant changes in surface composition [111]. Under low temperatures, it therefore seems that there is little difference between the choice of base used. It is, however, worth noting that these reaction systems are not guaranteed to preserve fibre quality. Additional experiments investigating the influence of temperature [91] and the concentration of OH^- ions [111] demonstrated that increasing either parameter led to a greater reduction in tensile properties.

Higher temperature systems may, however, facilitate a reduction in both the reaction time and the necessary concentration of KOH or NaOH. In doing so, a balance may be struck between these three variables, which optimises a recycling process for fibre quality. Where water was used in conjunction with 1 M phenol and 0.18 M KOH, a reaction temperature and time of 315 °C and 30 min resulted in fibres which demonstrated an equivalent tensile strength to virgin fibres [76]. A CFRP processed at 330 °C with 0.05 M KOH in butanol did result in a 5% reduction in strength and a 7% reduction in tensile modulus [105], while a propanol system with 0.09–0.36 M KOH resulted in a 5–11% reduction in strength in the 320–360 °C temperature range which was investigated [104]. Due to differences in carbon fibre type, solvent, and reaction conditions, it is not possible to determine what the exact influence of the KOH is and whether any reduction in fibre quality is exclusively due to the OH⁻ ions. It may be that it is the combination of heat, time, solvent, and the alkaline salt which together have a slightly negative impact on the properties of the fibres.

In addition to fibre strength, it is also pertinent to consider the surface characteristics of the recycled fibres. Mostly, under alkaline conditions, a slight increase in surface oxygen concentrations has been reported [76, 81, 91, 104], which is unsurprising as OH⁻ ions are likely to cause some surface oxidation. This may enable greater adhesion to a layer of sizing and/or a new matrix material, although this may be dependent on the specific polymer. As an example, some research has found that recycled carbon fibres displayed a reduction in interfacial shear strength (IFSS), and hence less adhesion to an epoxy droplet, when compared to virgin fibres. However, an increase in IFSS of 15.8% when using polypropylene was reported [105], thus demonstrating that the secondary application of the carbon fibres must also be considered when designing a suitable recycling process.

2.4.2 Glass fibres

Like carbon fibres, there is a wide variety of glass fibres used across different sectors. The impact of any recycling process on the quality of the recyclate is largely determined by the fibres' thermal durability and resistance to chemical attack. The major types used within composite materials include T-glass and E-glass fibres which are defined based on their chemical composition. The latter variety contains a greater proportion of Al_2O_3 and CaO, while T-glass fibres contain a larger quantity of SiO_2. This leads to an improved tolerance to both elevated temperatures and corrosion and, hence, is more likely to withstand chemical attack during a GFRP recycling process. For this reason, when discussing the impact of a recycling process on glass fibres, as in this section, it is essential to consider what the specific fibre types are.

2.4.2.1 Thermal and solvent effects

Where neat solvents are used, high temperatures are often necessary for the liberation of clean fibres; however, this does risk damaging them, especially in hydrolysis. Where water is present at subcritical conditions, there are likely to be hydronium (H_3O^+) ions generated. These ions attack the fibre surface, causing the major metal oxides to be leached from the fibre, only to be replaced by these hydronium ions. Therefore, the chemical composition of the glass fibre is a key factor in the degradation observed [112]. Hydrolytic systems in the temperature range of 280–350 °C have resulted in reducing the tensile strength of glass fibres by between 40% and 60% [65,

113, 114] where higher temperatures generally lead to greater fibre damage. As more H_3O^+ ions are generated at higher temperatures, it is, therefore, unsurprising that this effect has been seen in multiple works. Further characterisation of these processed fibres using scanning electron microscopy revealed that at the lower end of this temperature range, there was still a residual layer of the polymer matrix on the fibre surface [65, 114]. This is likely to have protected the fibre from further attack and/or sealed over any surface defects, which act as failure points during tensile tests. Further, the presence of this resinous layer may hinder adhesion to a new matrix material, which in turn limits potential secondary applications. Interestingly, this significant loss in tensile strength does not correlate with stiffness; in these examples, there was no significant change in Young's modulus [65, 114], and it has even been reported to increase [113] following subcritical hydrolysis.

In the absence of water, the mechanical properties of the glass fibres may be almost preserved. For example, after processing a glass fibre-reinforced epoxy with acetone in the temperature range of 260–280 °C, a strength reduction of 11–25% was observed [113]. As expected, the higher temperature did lead to weaker fibres, although, somewhat surprisingly, the stiffness was measured to increase. Amongst the most promising conditions reported in the academic literature is the supercritical methanolysis of a real end-of-life GFRP. A batch process at 275 °C, used in conjunction with 4-(dimethylamino pyridine) (DMAP), enabled the recovery of clean glass fibres which retained in excess of 93% of the strength of virgin material [115]. Other solvents may, however, have a much more detrimental effect. For example, the glycolysis of a glass-fibre-reinforced polyamide at 130 °C caused reductions in both strength and modulus of up to 45% and 55%, respectively [116].

More recently, there has been some work investigating the potential of other green solvents, although the efficacy of these is largely determined by the polymer matrix they are dissolving. D-limonene was able to decompose a glass fibre-reinforced polyester at both 300 °C and 390 °C. The recovered fibres were synthesised into a new composite, which retained only 85% and 64% of the strength of a fully virgin GFRP [117]. Clearly, it is desirable to minimise the reaction temperature, which may be more feasible with vitrimer-based materials. Vitrimers are a recently developed class of polymers which sit alongside thermosets and thermoplastics. They consist of covalently bonded cross-linked networks (like thermosets) but also contain a transesterification catalyst, such as zinc acetate. At elevated temperatures, this enables an exchangeable reaction to take place which releases long polymer chains from their rigid positions and hence results in the material acting more like a thermoplastic [118]. With such materials, relatively low temperatures can be used to recover clean fibres. One example involves the application of diethylene triamine (DETA) and xylene, which enabled degradation of the vitrimer matrix in 12 h at 75 °C. The glass fibres recovered were used to manufacture a secondary composite, which displayed a similar flexural strength to the fully virgin samples following one and two recycling loops. A third recycling loop did, however, present a loss in strength of ~8% [119].

These examples illustrate that, in order to recover clean fibres, the solvent, time, and temperature must be matched to the specific polymer matrix. However, there is often a severe impact on the quality of the glass fibres which is largely a consequence of thermal and chemical degradation. To avoid this, acidic systems have been explored in the academic literature, as discussed in the following section.

2.4.2.2 Effect of acids

Typically, strong acids such as sulphuric and nitric acid are necessary to break down the polymer matrix. These strong acids are frequently supplied at very high concentrations. For example, 98% sulphuric acid (H_2SO_4) has been used to solubilise an epoxy resin and liberate glass fibres. Long reaction times of 50 h were necessary, but as the process was successful at room temperature, the glass fibres experienced very little change in modulus, with a reduction from 74 GPa to 72 GPa. There was, however, a significant change in strength, with the manufacturer's data reporting 3.34 GPa and the researchers measuring 1.03 GPa post-recycling. This was attributed to a loss of the sizing coating the material, and some strength may be recoverable if a new sizing layer is applied [120]. Other strong acids, such as nitric acid, have also been investigated for their suitability in recycling GFRPs. In one example, a 4 M solution completely solubilised a bisphenol F epoxy resin after a very long reaction time of 400 h at 80 °C [101]. However, after being exposed to this solution for just 100 h, there was a 30% mass loss from E-glass fibres, while T-glass fibres were unchanged. This suggests that the E-glass fibres are more susceptible to chemical corrosion [101]. Additional work applied 6 M nitric acid to a T-glass fibre composite. Clean fibres with a reduction in tensile strength of just 3.5% were recovered after processing at 70 °C for 250 h, although it is worth noting that increasing acid concentration and reaction temperature to drive a faster reaction rate resulted in strength reductions of up to 15% [121]. This further illustrates that while E-glass fibres cannot withstand acidic chemical recycling systems, T-glass fibres can and, therefore, it is essential to understand the feedstock material when designing a chemical recycling process.

Although it is dependent on the polymer, it may also be possible to avoid the use of these strong, harmful acids through elevated temperatures. For example, an acetic acid/$AlCl_3$ system (which is considered a weak organic acid and a weak Lewis acid, respectively) enabled the degradation of an unsaturated polyester in 9 h at 180 °C. The fibres recovered retained more than 96% of the strength of the virgin fibres which suggests that, while relatively slow compared to higher-temperature systems, this may be a promising route for recycling glass fibres [96]. However, it is worth noting that the fibre type was not specified, and so, as described above, it may not be suitable for E-glass if T-glass fibres were the material investigated in this work.

2.4.2.3 Effect of bases

The influence of basic media on the properties of glass fibres following solvolysis of their composites is less widely reported than the effect of acids. There are however, some published research which illustrates severe detrimental effects no matter the solvent or operating conditions. Most commonly, both KOH and NaOH are used as sources of OH⁻ ions, with both chemically etching the fibre surface. This has been shown to occur even at short reaction times of 5 min and low concentrations of 0.2 M [114]. In this work, virgin fibre strength was 2.14 GPa; when processed with water alone at 350 °C for 5 min, it was reduced to 1.24 GPa, and when NaOH was incorporated at the same conditions, the strength was further reduced to 0.69 GPa. However, it is worth noting that there was a difference in resin removal of 15% between the two samples, and this larger strength reduction in the presence of NaOH may be due to a greater degree of resin removal. Similar results at the same process conditions have also been reported elsewhere [65].

In addition to this loss in strength, it has also been noted that under hot alkaline conditions, there is a significant risk of mass loss from the glass fibres. This is due to the solubility of glass in highly alkaline solutions, where the SiO_2 forms silicate salts. For example, in one system using 1 M KOH at 150 °C, a 10 wt.% loss in mass was observed specifically due to the degradation of the fibres themselves [122]. A more comprehensive study investigated the effects of both KOH and NaOH at concentrations in the range of 0.5–3 M, temperatures from 75 to 95 °C, and treatment times from 1 to 5 h, and measured the corresponding effect on fibre diameter [123]. The results showed that increasing any of the parameters led to a greater reduction in the fibre diameter as the glass itself is degraded. At the most severe conditions (95 °C, 5 h, 3 M NaOH), there was a reduction in the fibre diameter of ~80%, which is thought to have a corresponding impact on the fibre strength [123]. Interestingly, NaOH was more corrosive towards glass than KOH, and so this should also be considered in the selection of the parameters for a GFRP chemical recycling process.

2.5 Composite additives and influence on recycling

By their very nature, FRPs consist of multiple different types of material. This book mostly considers the two main constituents: the fibre reinforcement and the binding matrix. In addition to these, there are often additional materials, such as coatings or fillers like bulking agents, which must also be considered when discussing the end-of-life options for an FRP.

2.5.1 Coatings and paints

Most academic research into the influence of coatings on organic degradation products has only considered a thermal, rather than chemical, recycling process. Nevertheless, some parallels may still be drawn, especially where HTP conditions are used. The two types of coatings considered in this section are protective polymer films and paints, both of which yield different classes of chemical products.

Paints are often used to coat the outside of CFRPs, either for decorative purposes or to protect the surface of the composite material. Although the paint film will typically make up a low proportion of the mass of composite waste, it is still pertinent to consider what organic products they may decompose into, and thus, it is relevant to identify their initial composition. Paints typically consist of five main components [124]:

1. Solvent – Provides a medium to disperse the other components and is used as a rheology modifier.
2. Pigments – Provide the final colour and opacity of the paint.
3. Extenders (or fillers) – Larger pigment particles which are added to strengthen the paint film, lower cost, and improve adhesion.
4. Binder – Usually a polymer and used to form a matrix which holds the pigments in place after the paint has dried.
5. Additional additives – Can include dispersants, silicones, driers, anti-settling agents, bactericides, or fungicides. These are added to influence the final properties of the liquid or dried paint.

The solvent used in paints may be water or a mixture of organics, usually classified as volatile organic compounds (VOCs). These are largely removed and released into the atmosphere as the paint dries and are therefore not thought to significantly influence the product distribution from an FRP recycling process. If any residual solvent remains in the paint, it will likely be released during the recycling process. For water-based paints, this may result in a slight increase in the moisture content of a furnace or reactor, which may subsequently take part in additional reactions. The VOCs used in oil-based paints are typically benzene and alkylated derivatives [125], which are already common products from the decomposition of the resin. For this reason, it is also thought that this won't significantly alter the product distribution from an FRP recycling process.

Pigments may be classified as organic or inorganic. In the latter case, the majority of pigments are formed from various metal oxides, with different blends providing specific colours. Amongst the most common is TiO_2 which is reportedly present in ~70% of commercial and industrial paints [124]. This is a white powder and provides a glossy finish to the final coat. Other common components include iron oxides, zinc oxides, and carbon black. If present during a chemical recycling process, it is unlikely that the chemical structure of these pigments will change significantly; the only change may be additional oxidation. In the case of carbon black, this may lead to a higher concentration of CO_2 and a subsequent reduction in the GCV of any recovered gas. There is a risk that these metal oxides may contaminate the surface of the fibres, and, if present in the recoverable liquid fraction, may need to be removed. The presence of these transition metals may catalyse the decomposition reaction; however, it is not thought that their concentration will be high enough to significantly influence the recycling process. Organic pigments are generally complex molecules, such as phthalocyanines, which co-ordinate with metal ions, for example copper, resulting in a bright blue or green colour. These are cyclic structures with a moderate concentration of nitrogen [126] and are therefore likely decomposed into similar products as the resinous matrix.

Extenders or fillers are typically minerals such as magnesium silicate, mica, or kaolin. When heated in solvolysis processes, these materials are likely to carbonise forming a range of different carbonates [127]. Similar to the inorganic pigments, these substances may need to be considered in order to preserve fibre quality or be precipitated from the liquids recovered.

Binders present in paint formulations are likely to contribute to the potentially useful organic products which may be recovered from the thermal or chemical recycling of FRPs. These are generally resins based on acrylic or epoxy structures [124], as are common polymer matrices used in composite materials. The organic products generated following the decomposition of these binders are, therefore, likely to be similar to the products recovered from the matrix. Any additional additives are generally present in very small quantities, often constituting less than 1% of the paint [124]. As the paint itself also only makes up a small quantity of a CFRP by mass, these additional additives are not thought to significantly influence the organic products recovered.

2.5.2 Fillers

Filler materials may be included into the resinous matrix to aid processing or provide specific properties to the final product. Conveniently, they also often lower the overall cost of a CFRP, as they utilize inexpensive materials such as calcium carbonate, kaolin, alumina trihydrate, and calcium sulphate.

Kaolin is a hydrated aluminium silicate, and while there are no reported studies on the influence of kaolin on the organic products recovered from composite materials, it is possible to consider its own decomposition products. As a thermally stable material, only the loss of moisture has been reported when exposed to temperatures up to 500 °C [128]. For this reason, it is likely that aluminium silicate can be recovered if a solvolysis recycling process is used. Other than a possible increase in water content, its presence is also unlikely to significantly influence the mixture of organic products.

Calcium sulphate ($CaSO_4$) is a relatively inert material, only decomposing to CaO and liberating SO_2 gas at high temperatures in excess of 1,350 °C [129]. For this reason, this filler material is unlikely to influence the organic products recovered from an FRP chemical recycling process. If $CaSO_4$ is used in large quantities as a filler, however, its separation from the liquid product or removal from the fibre surface will need to be considered. It is also worth noting that it can exist as a hydrate ($CaSO_4 \cdot 2H_2O$) and, depending on the form used in a composite, it may alter the recycling process slightly. The hydrated sulphate begins to lose moisture at 120 °C, with only the anhydrous form existing at temperatures above 190 °C [130]. Therefore, the presence of $CaSO_4 \cdot 2H_2O$ in an FRP will likely contribute to the moisture content of the condensable product obtained from pyrolysis processes. In solvolysis, any water molecules present may recombine with the anhydrous $CaSO_4$ upon cooling the reactor down to ambient conditions.

2.6 Environmental and economic considerations

The environmental benefits of recycling composites were reported in 1997 by Hedlund-Åström, when they reported a life cycle assessment (LCA) of a mechanical recycling process. The LCA is often used alongside a life cycle cost (LCC) analyses to assess the environmental and economics of recycling methods and is referred to as an integrated Life Cycle Engineering approach. Besides the LCC, a techno-economic assessment (TEA) can be used, which evaluates both the technical feasibility and economic viability of a recycling technology.

The LCA should adopt a framework as specified by the International Organisation of Standardisation (ISO) based on standards ISO 14040 2006 and ISO 14044 2006. Four categories are identified: (1) goal and scope definition, (2) life cycle inventory analysis, (3) impact assessment, and (4) interpretation of results. When assessing the impacts of a recycling process, the outputs from an LCA are subject to the inventory analysis inputs and the assumptions made by the practitioner. The procedures to determine LCA for fibre-reinforced plastics using three case studies were reported by Gharde and Kandasubramanian [131]. Although solvolysis is not considered here, this paper provides an excellent overview of the approach.

The literature has numerous case reports of LCAs and TEAs to assess the viability of solvolysis recycling, and has gathered momentum with several articles appearing in the last few years as the popularity of solvolysis increases. Across the reports, the assessments generally show favourable results, with a number of authors highlighting the superior environmental and economic benefits of solvent-based recycling [132, 133] demonstrating that solvolysis is worth pursuing as a recycling technology.

Most articles for LCA are for CFRPs with little on GFRPs. This may demonstrate the suitability of solvolysis to recycle CFRPs, or perhaps illustrates a research gap that needs further investigation. The LCA is invariably undertaken at the lab-scale, and it is anticipated that electricity consumption and, therefore, the overall environmental impact will decrease per kilogram of material processed upon potential scale-up.

2.6.1 Life-cycle analysis of CFRP solvolysis

2.6.1.1 Lab-scale processes

Poranek et al. [134] used LCA to assess the viability of two lab-scale solvolysis approaches to recycle carbon fibre: one based on EG and potassium hydroxide at ambient pressure, and the other a plasma-enhanced nitric acid solvolysis. They show that the former approach has environmental benefits, whereas the latter approach has energy-associated issues that should be addressed through further development and optimisation. This demonstrates the use of undertaking LCA even for low TRL processes. A significant paper by Kamali et al. [135] collated 28 articles to harmonise the data across recycling processes where LCA was considered. They showed that supercritical hydrolysis has the lowest environmental impact, while energy-autonomous pyrolysis showed negative GHG emissions and produces fibres with 80% tensile strength compared to virgin fibres. Vogiantzi and Tserpes (2025) [136] compared four recycling approaches and concluded that high voltage fragmentation (HVF) was the most environmentally friendly process for carbon fibre recovery, with solvolysis and pyrolysis requiring large energy inputs. Solvolysis was also reported to have higher costs but produced high-quality fibres. The literature does report instances where pyrolysis performs better than solvolysis across environmental and health-based impacts for CFRP recycling [137, 138]. In these cases, supercritical water is used as the solvolysis route. However, caution should be taken here as sub- and super-critical water is often not the most favourable solvolysis conditions. In other cases, the advantages of solvolysis over pyrolysis are reported [139]. Combining water with organic solvents to give binary mixtures can lead to lower production costs, environmental and human health impacts compared to water alone [137]. The choice of solvent is an important criterion when assessing the viability of the solvolysis process, with solvents requiring less process severity showing greater environmental and economic benefits.

Supercritical water was used in the recovery of fibres from CFRP and had a lower environmental impact compared to landfilling [140], achieving an environmental gain of about 80% [141]. The process also gives fibres with excellent mechanical properties. Chatzipanagiotou et al. [142] compared the LCA of supercritical water solvolysis and plasma-enhanced solvolysis to that of virgin carbon fibre for use in additive manufacturing. They showed that the solvolysis processes, even though at the lab-scale, gave a lower environmental impact across most of the assessment categories compared to virgin fibre.

Kooduvalli et al. analysed the embodied energy of several pyrolysis and solvolysis methods, basing their study on the cradle-to-grave of 1 kg CF-epoxy laminates. Four solvolysis approaches were considered, based on supercritical conditions in a batch reactor using water with catalyst and solvent additives. Both pyrolysis and solvolysis showed that fibres can be recovered having lower embodied energy than virgin fibres with pyrolysis coming out on top [143]. This emphasises the earlier comment that water is not always the best option and that other solvents and catalysts will lessen the embodied energy required for solvolysis.

2.6.1.2 Large-scale processes

The ground-breaking work of Kawajiri and Kobayashi is the first to compare the LCA of production-scale solvolysis and pyrolysis based on the pyrolysis process of Carbon Fibre Recycling Co. and the solvolysis process of Hitachi Chemicals [144]. The Hitachi Chemicals process uses benzyl alcohol with 10 wt.% of K_3PO_4 soaked at 190 °C for 2 h followed by a water rinse and heat treatment at 350 °C for 1 h, whereas the Carbon Fibre Recycling Co. uses pyrolysis at 550 °C for 3.7 h. It should be noted that the solvolysis production scale was smaller to that of the pyrolysis and was scaled accordingly to match the pyrolysis. The results indicated that both techniques led to a reduction in environmental impacts compared to virgin CF, with pyrolysis giving the lowest value of 1.52 kg-CO_2 eq kg^{-1} compared to 1.92 kg-CO_2 eq kg^{-1} for the solvolysis process; however, the tensile strength of the fibres from the solvolysis was better than those from the pyrolysis (3,200 MPa compared to 2,000 MPa, respectively). The authors state, however, that assumptions and uncertainties in the data should be recognised and considered [144].

    Urruzola et al. describe a conceptualised combined mechanical, pyrolysis, and solvolysis large-scale plant for carbon fibre recycling. The solvolysis used water, acetic acid, and hydrogen peroxide for the removal of the polymeric matrix. The authors concluded that the recovered carbon fibre is better from an environmental efficiency perspective and is 35% less expensive than virgin carbon fibre [145].

2.6.1.3 Case studies

Alavi et al. investigated four recycling routes (landfilling, mechanical recycling, pyrolysis, and solvolysis) for carbon fibre wind turbine blades in Australia [146] and showed that solvolysis using acetic acid and sodium hydroxide was the most environmentally effective method, with pyrolysis a close second. A factor worth considering was that the CO_2 reduction was state-specific and related to the amount of installed wind capacity and, therefore, the amount of waste available to process.

    Merlo-Camuñas et al. [147] reported the LCA for replacing two parts used in trains with those made from recycled carbon fibre. A three-step process was proposed that first used low-temperature pyrolysis at 450 °C, followed by a grinding step, and finally an acetic acid/hydrogen peroxide step to remove the resin. Although this approach is not solely based on solvolysis, it demonstrates the advantages of combining solvolysis with other approaches and the potential of using recycled carbon fibre in lightweighting trains.

Wu et al. [148] developed an integrated LCA and LCC framework for substituting an aluminium aircraft door with one made from recycled carbon fibre. Acetic acid and sodium hydroxide solutions were used for the solvolysis. They showed that solvolysis gave the best GHG emission performance over pyrolysis and mechanical recycling; however, manufacturing the composite door led to increased costs over the metallic one.

2.6.2 Life-cycle analysis of GFRP solvolysis

As mentioned above, there are few cases of LCA on the recycling of GFRP. Sobek et al. [149] used an oxidative liquefaction method to recover glass fibres from wind turbine blades as well as chemicals from the epoxy resin matrix. The composite was degraded under hydrothermal conditions with peroxide at 50–350 °C, a residence time of 30–90 min, and between pressures of 20–40 bar. High-concentration peroxide (40 wt.%) gave the best quality glass fibres but resulted in energy use greater than that of virgin glass fibres; however, replacing the oxidant with a theoretical zero-input oxidant led to a lower net climate change impact. A low climate change impact was seen for lower peroxide concentration (15 wt.%); however, low-quality fibres were obtained. Further research into identifying zero-input oxidants, perhaps from waste sources, was recommended. However, it was reported that solvolysis with oxidation has high operational costs and leads to a reduction in the mechanical and physical properties of the fibres, limiting their application and is therefore unprofitable [150].

2.6.3 Economic feasibility of chemical recycling

The cost-benefit of solvolysis over several recycling approaches, including mechanical, pyrolysis, electrochemical, incineration, and landfill, was reported by Wei and Hadigheh [150]. Acid solvolysis and alkali solvolysis methods were considered. Acid solvolysis required high capital investment; however, changing the acid to an alkali reduced the investment. Overall, solvolysis may give the highest profit in comparison with other recycling methods for the recovery of fibres. Thermal recycling methods required the lowest energy input, while solvolysis and electrochemical methods have the lowest global warming potential. The cost-benefit was considered for both GFRP and CFRP and showed, as expected, that carbon fibre recycling is more profitable due to the low production cost of glass fibre.

Rosa et al. [151, 152] reported the LCA and cost-benefit analysis of using Connora Technologies' cleavable thermoset formulations to recycle and recover carbon fibre-reinforced thermosets and their reuse in epoxy thermoplastics, which were assimilated to polycarbonate. An acetic acid/sodium hydroxide solution was used for the solvolysis process. For closed-loop recycling of the recovered carbon fibres, costs compared to virgin carbon fibres were seen along with a reduction in global warming potential. For open-loop recycling, where recovered short carbon fibres and the thermoplastic epoxy were considered, beneficial economic and environmental benefits were seen with reductions in CO_2 emissions of -97 kg CO_2eq for the thermoplastic epoxy and -128 kg CO_2eq for the carbon fibres.

In many cases, solvolysis is more cost-effective for reclaiming fibre from CFRP waste than other recycling methods. Nonetheless, the economic and environmental viability of using solvolysis is a vitally important factor when considering the commercial scale-up of composite recycling. Tapper et al. [153] suggest that recycling technologies should produce high-quality materials that can be remanufactured into new materials to gain the full benefit of emission savings and economic value.

For recovered carbon fibres, their mechanical performance-to-price ratio should be considered and needs to be higher than that of glass fibre to be attractive to the market [141]. High recycling capacities and high carbon fibre recovery rates are required for solvolysis processes to overcome both the price of virgin fibre and recycled fibre from cheaper techniques. Indeed, recycled fibres from supercritical water solvolysis are not competitive in the recycled glass fibre market due to the very high treatment cost (over 3.5 € kg^{-1} of fibre) even at high capacities of 4,000 t year^{-1} [154].

2.7 Challenges and outlook

2.7.1 Challenges

As this chapter has discussed, there is a wide selection of different chemical recycling techniques which have been applied to FRPs. Unfortunately, the vast majority of these remain at the lab scale, with few examples capable of processing more than several grams of material at a time, and even fewer reaching commercial maturity. At present, there are many challenges which must be addressed before solvolysis-based FRP recycling processes become commonplace. Chief amongst these are the economic conditions currently present in both the FRP manufacturing and recycling sectors. As alluded to above, many recycling processes are relatively expensive compared to the financial cost of making virgin material. This is particularly prevalent for GFRPs, where glass fibres are exceptionally cheap to produce at about 0.75–3 USD kg^{-1} [13]. Both technological and policy-driven approaches could be taken to mitigate this. Chemical recycling with solvents has an advantage over pyrolysis through the generation of organic, even value-added, products from the degradation of the resin. A summary of the products typically gained, along with upgrading techniques, has been published elsewhere [13]. This could allow the development of additional revenue streams, which helps mitigate the high cost of chemical processing, although there will likely be additional costs to consider for any necessary downstream processing. From a policy or legislative perspective, there are two primary levers which could be used. Landfill taxes, which are widely variable around the world, could be levied which makes the disposal cost of waste composites higher than their recycling costs. Secondly, additional levies could be placed on original equipment manufacturers (OEMs) who use more than a predetermined quantity of fully virgin material in their products, thus stimulating market growth of secondary composites. A similar policy, known in the UK as the Plastic Packaging Tax, has been effective in encouraging packaging manufacturers to incorporate a minimum of 30% recycled plastic in their products [155].

However, at present, there is an unproven market for secondary composites, and the supply chain is at its infancy. Concerns regarding fibre quality, and hence the performance of recycled materials, are valid; although processes exist which claim to produce fibres with properties equivalent to virgin fibres, they are likely to degrade over multiple recycling loops. The ability of fibres to withstand these recycling processes is also dependent on the fibre type, and hence the traceability of what specific fibre is used in real end-of-life parts must be established, along with sufficient quality control measures. In high-performance applications where safety is critical, particularly in aircraft, it is likely that recycled materials will never be used for key structural components. There may, however, be greater scope to use them in non-structural parts like seat frames or wall panelling.

The lack of traceability regarding fibre reinforcement extends to the traceability of an entire composite part, and this too represents a significant challenge. After manufacturing, many FRP components are fitted into a final product; the product is used and then disposed of. At the disposal stage, which may occur many years later, there is no easy way to determine the fibre type, the resin, or the presence of any fillers or additives. As described in this chapter, the recycling process chosen is often dependent on the specific composite to be recycled. However, in practice, a commercial process must be capable of handling a mixed feedstock. This means that it is likely to use harsh conditions and generate a complex mixture of potentially harmful organic products. To avoid this, two possible solutions may be possible. An effective sorting process, perhaps using recent innovations in infrared detection, may at least be able to separate out a mixed feedstock based on resin type. This has been trialled by Panasonic Electric Works [156]. Secondly, partnerships between manufacturers, end-users, and recycling companies might also be possible which would allow materials to return to OEMs at the end of their first useful life. This could be achievable through the application of blockchain technology, as recently demonstrated by a collaboration between Japanese companies Fujitsu and Teijin [157]. However, to achieve this, there would have to be suitable economic incentives. If delivering recycled materials is not cheaper than manufacturing virgin material, such a system is likely to rely on the introduction of government legislation.

In summary, the primary challenges facing commercial chemical recycling processes for FRPs are the relatively high financial costs (both initial capital and operating), an exceptionally diverse range of different materials which will require different process conditions, and a lack of a proven market for secondary composites. Despite these barriers, there are a limited number of cases where the chemical recycling of FRPs is being scaled commercially, as discussed in the following section.

2.7.2 Commercial processes

Although in its infancy, there are some examples of commercial solvolysis processes for the recycling of CFRPs. Adherent Technologies Inc. (ATI), based in New Mexico, USA, was one of the early developers of industrial-scale solvolysis for composites. They use a wet chemical method, termed the Jumbo process, to recover clean carbon fibres which retain more than 95% of the strength of virgin material [158].

In the UK, Uplift360 has developed a chemical process capable of dissolving thermoset resins at room temperature and pressure. The solvent is recovered and used through multiple recycling loops; however, beyond this, there are few published details about the process [159]. A patent filed by the company provides some additional information regarding the Uplift360 technology: ozone, dissolved in water, forms hydroxyl radicals when exposed to UV radiation. These radicals then attach to linkage sites within the polymer matrix. This subsequently degrades the matrix and liberates clean fibres [160]. Announced in 2025, the company has received funding to scale the existing process and develop an integrated pilot line, although target throughputs have not been mentioned [161].

The patented Hitachi Chemical system described above (benzyl alcohol, 10 wt.% K_3PO_4 at 190 °C for 2 h, followed by a 350 °C heat treatment) also remains at the pilot scale, with a 2×200 L baths capable of processing ~12 tonnes of CFRP per year [162].

As these examples remain at the pilot scale, they are not as developed as other pyrolysis-based FRP recycling technologies. Gen2Carbon (UK), CFK Valley Recycling GmbH (Germany), Mitsubishi Chemical (Japan), and Carbon Rivers (USA) all operate commercial FRP recycling systems at much greater volumes than these pilot-scale solvolysis systems. Crossing the barrier from a TRL of about 4 to above 7 is likely to remain difficult in the coming years due to the previously discussed challenges. There remains, however, an opportunity for solvolysis to generate additional revenue through the recovery of organic products, deliver energy savings compared to pyrolysis, and hence achieve an economically viable recycling system.

2.7.3 Achieving a circular economy for composites

Notwithstanding these recent examples of commercialisation, there remain some additional approaches which could be harnessed to achieve a circular economy for composites. Given the large volume of materials currently on the market, there must be greater consideration of utilising not just the fibres, but the organic products recovered from the matrix. At present, these are largely considered waste, yet they may be upgraded to realise additional economic value as well as improving circularity. In addition to this, there are recent innovations in materials design which, together with chemical recycling, can support a circular economy.

The first of these is designing polymers for recyclability. This has been evidenced with the commercialisation of Elium resins by Arkema, a high-performance methyl methacrylate polymer with properties similar to those of thermosets [163]. Academic work has demonstrated that solvolysis of this resin can recover usable monomers, which may be directly incorporated into a new composite [164]. This was completed with production waste, where the recycled monomers displaced up to 7.5% of virgin material which resulted in an increase in stiffness, shear strength, bending properties, and thermal stability [164]. This design-for-recyclability approach has also been extended to thermosets. These exploit new lockable/unlockable chemistries, where initiators "lock" polymer chains together, which can later be "unlocked" under specific conditions [165]. Aditya Birla Chemicals currently produces "Recyclamine"; a new class of amine-based curing agent. Originally developed by Connora Technologies, this agent cross-links molecules of DGEBA together but, unlike other hardeners, contains acid-cleavable functional groups [166]. The medium frequently used is a 25 vol.% acetic acid solution, heated to about 80–100 °C, which causes the cleavable bonds to be broken within 1–6 h. Subsequent neutralisation precipitates an epoxy thermoplastic, which may be concentrated using a centrifuge or rotary evaporator [152, 167, 168]. Although the recovered polymer was not suitable for use in a new thermoset material, it was successfully incorporated into a thermoplastic composite. This demonstrated slightly lower tensile strength than a virgin counterpart, although this was attributed to less aligned fibres rather than a lower performance of the resin [167]. Additional innovations in this area include those of Swancor in Taiwan, who have developed a recyclable and reusable epoxy called EzCiclo. This can be recycled by their CleaVER technology (a liquid system) agitated at 130 °C for 3 h to yield glass or carbon fibres and monomers and oligomers. The monomers and oligomers can be reused in the manufacture of new EzCiclo resin, with the CleaVER liquid reused in the resin itself [169].

Alongside designing polymer matrices for recycling, the use of bio-based resins is also gaining interest across academic and commercial spheres. Moving from fossil-based to bio-based feedstocks is a key pillar of achieving a circular economy, and this principle can also be applied to further achieving sustainability in FRPs. One example is the Super Sap technology developed by Entropy Resins. This contains epoxidised pine oils along with fossil-derived BPA/BPF-based polymers. When cross-linked with Recyclamine, this can be chemically recycled; however, the initial pine oil content was not disclosed nor fully separated. Instead, only an organic mixture was recovered, which may be used as a new matrix material [170]. Alternative bio-based feedstocks which may be suitable for use in FRPs include epoxidised vegetable oils, unsaturated polyesters from bio-based diols, and phenolic-based resins from lignin [171]. The development of such polymers remains an active research area, and as the chemical structure of these alternatives is similar to fossil-derived counterparts, it is possible that existing chemical recycling processes can be used as an effective waste management strategy.

2.8 Conclusions

This chapter has outlined the reaction systems, process conditions, and underlying mechanisms which underpin the chemical recycling of both thermoplastic and thermoset-based composites. Although mechanical and thermal recycling routes are already industrially established, solvolytic systems provide a critical advantage: the ability to recover near-virgin quality fibres and valuable organic chemicals. Recycling both components and maintaining their value not only improves circularity but also increases supply security and may reduce environmental impact.

The recycling of thermoplastics is generally easier than that of thermosets; the long polymer chains are mobile at elevated temperatures, and original monomers are recoverable. Hydrolysis, alcoholysis, and glycolysis have all been shown to selectively cleave polymer backbones under a range of conditions, producing monomers and oligomers with high recovery yields. Due to the cross-linked and infusible nature of thermosets, the solvolysis of these composite materials is challenging. However, considerable progress has been achieved since the early 2000s, with hydrolysis, alcoholysis, and aminolysis all capable of decomposing complex resin networks. Hydrolysis, whether catalysed by acids, bases, or metal salts, remains the most extensively studied route for thermosets. Acidic systems, such as nitric and acetic acid, are particularly effective at breaking down epoxy and polyester matrices, though care must be taken to avoid excessive fibre degradation. Metal chlorides, such as $ZnCl_2$, $AlCl_3$, and $MgCl_2$, have emerged as versatile catalysts, providing selective cleavage of C–N bonds and lowering activation energies. Meanwhile, novel green systems based on deep eutectic solvents demonstrate the potential to replace traditional solvents and deliver efficient degradation at temperatures of less than 200 °C.

The preservation of fibre properties is central to developing a solvolysis recycling process. Carbon fibres generally withstand a range of temperatures, pressures, times, solvents, and additives with minimal deterioration in tensile strength or modulus, although all of these parameters must be well controlled. Surface oxidation introduced during these treatments may even enhance adhesion to the polymer matrix in a secondary composite. Glass fibres, particularly E-glass fibres are less thermally stable. As such, only T-glass fibres can be recovered with near-virgin properties under carefully optimised systems. However, their low intrinsic value and the relative cost of recycling mean that commercial interest here is limited. Emerging techniques for maintaining fibre length or realigning them represent further opportunities for retaining mechanical performance.

Beyond these technical achievements, the environmental and economic implications of solvolysis have been increasingly examined through life cycle and techno-economic assessments. Studies consistently demonstrated that solvolysis offers substantial environmental benefits compared to landfilling or incineration; there are significant reductions in greenhouse gas emissions and energy demand relative to virgin composite production. Pyrolysis may offer slightly lower emissions than solvolysis; however, the environmental impact is dependent on the specific scenario, feedstock materials, and process conditions. Economically, solvolysis can outperform other recycling methods, particularly for CFRPs, where the fibre component is particularly valuable. It may offer superior profitability to pyrolysis, especially when mild alkaline systems are used or when recyclable-by-design approaches are implemented. However, scalability and solvent recovery are critical determinants of long-term viability. Although solvolysis is not yet cost-competitive with pyrolysis for low-value glass fibres, some research has highlighted the potential for environmental improvements due to lower process temperatures.

As countries around the world strive towards net zero, there is a growing need to consider the sustainability of composites. To this end, the focus is on improving the recyclability of both thermoplastic and thermoset resins, whilst also considering bio-based feedstocks. Achieving a fully circular economy for composite materials will depend on parallel advances in guaranteeing recycled fibre quality, developing solvent recovery infrastructure, and creating stable market demand for secondary fibres and organic products. When combined with rigorous LCA and TEA evaluations, solvolysis offers a clear pathway toward sustainable composite manufacturing by closing the loop between production, use, and end-of-life recovery.

References

[1]   C. B. Farinha, J. De brito, and R. Veiga, "Assessment of glass fibre reinforced polymer waste reuse as filler in mortars", *J. Clean. Prod.*, vol. 210, pp. 1579–1594, Feb. 2019, doi: 10.1016/J.JCLEPRO.2018.11.080.

[2]   S. Karuppannan Gopalraj, and T. Kärki, "A review on the recycling of waste carbon fibre/glass fibre-reinforced composites: fibre recovery, properties and life-cycle analysis," *SN Appl. Sci.*, vol. 2, no. 3, pp. 1–21, 2020, doi: 10.1007/s42452-020-2195-4.

[3]   S. S. Yao, F. L. Jin, K. Y. Rhee, D. Hui, and S. J. Park, "Recent advances in carbon-fibre-reinforced thermoplastic composites: A review", *Compos. B Eng.*, vol. 142, pp. 241–250, 2018, doi: 10.1016/j. compositesb.2017.12.007.

[4]   G. Oliveux, L. O. Dandy, and G. A. Leeke, "Current status of recycling of fibre reinforced polymers: Review of technologies, reuse and resulting properties", *Prog. Mater. Sci.*, vol. 72, pp. 61–99, 2015, doi: 10.1016/j.pmatsci.2015.01.004.

[5]   M. Agrawal, and R. T. D. Prabhakaran, "Effect of fibre sizing on Mechanical properties of carbon reinforced composites: A Review," *Org. Polym. Mater. Res.*, vol. 1, no. 2, pp. 1–5, 2020, doi: 10.30564/ opmr.v1i2.1683.

[6]   L. Mao, H. Shen, W. Han, L. Chen, J. Li, and Y. Tang, "Hybrid polyurethane and silane sized carbon fibre/epoxy composites with enhanced impact resistance", *Compos. Part A Appl. Sci. Manuf.*, vol. 118, pp. 49–56, Mar. 2019, doi: 10.1016/J.COMPOSITESA.2018.12.014.

[7]   M. Shekarchi, E. M. Farahani, M. Yekrangnia, and T. Ozbakkaloglu, "Mechanical strength of CFRP and GFRP composites filled with APP fire retardant powder exposed to elevated temperature," *Fire Saf. J.*, vol. 115, no. July, pp. 103178, 2020, doi: 10.1016/j.firesaf.2020.103178.

[8]   M. Monoranu, R. L. Mitchell, K. Kerrigan, J. P. A. Fairclough, and H. Ghadbeigi, "The effect of particle reinforcements on chip formation and machining induced damage of modified epoxy carbon fibre reinforced polymers (CFRPs)", *Compos. Part A Appl. Sci. Manuf.*, vol. 154, pp. 1–14, 2022, doi: 10.1016/j. compositesa.2021.106793.

[9]   Y. Yan, *et al.*, Non-Destructive Testing of Composite Fibre Materials With Hyperspectral Imaging – Evaluative Studies in the EU H2020 FibreEUse Project, *IEEE Trans. Instrum. Meas.*, vol. 71, pp. 1–12, 2022, doi: 10.1109/TIM.2022.3155745.

[10]   A. Faltynkova, G. Johnsen, and M. Wagner, "Hyperspectral imaging as an emerging tool to analyze microplastics: A systematic review and recommendations for future development," *Microplast. Nanoplast.*, vol. 1, no. 1, pp. 1–19, 2021, doi: 10.1186/s43591-021-00014-y.

[11]   M. Keith, "Recycling high performance carbon fibre reinforced polymers using sub- and supercritical fluids," University of Birmingham, Birmingham, 2019. Accessed: Nov. 12, 2024. [Online]. Available: https://etheses.bham.ac.uk//id/eprint/10111/

[12]   G. Oliveux, L. O. Dandy, and G. A. Leeke, "Current status of recycling of fibre reinforced polymers: Review of technologies, reuse and resulting properties", *Prog. Mater. Sci.*, vol. 72, pp. 61–99, Jul. 2015, doi: 10.1016/j.pmatsci.2015.01.004.

[13]   M. J. Keith, B. Al-Duri, T. O. McDonald, and G. A. Leeke, "Solvent-Based Recycling as a Waste Management Strategy for Fibre-Reinforced Polymers: Current State of the Art," *Polymers (Basel)*, vol. 17, no. 7, pp. 843, Mar. 2025, doi: 10.3390/polym17070843.

[14]   G. Oliveux, J. L. Bailleul, A. Gillet, O. Mantaux, and G. A. Leeke, "Recovery and reuse of discontinuous carbon fibres by solvolysis: Realignment and properties of remanufactured materials", *Compos. Sci. Technol.*, vol. 139, pp. 99–108, 2017, doi: 10.1016/j.compscitech.2016.11.001.

[15]   JEC, "New production line for Lineat Composites with Nigel Walker as Technical adviser," *JEC Magazine*, 2025. Accessed: Sep. 11, 2025. [Online]. Available: https://www.jeccomposites.com/news/ by-jec/lineat-composites-to-launch-new-production-line-with-nigel-walker-as-technical-adviser/? news_type=announcement,process-manufacturing&end_use_application=aerospace,sports-leisure-recreation&process=fibre-placement&tax_product=carbon-fibre

[16]   C. Branfoot, H. Folkvord, M. Keith, and G. A. Leeke, "Recovery of chemical recyclates from fibre-reinforced composites: A review of progress", *Polym. Degrad. Stab.*, vol. 215, pp. 110447, Sep. 2023, doi: 10.1016/j.polymdegradstab.2023.110447.

[17]   V. Sinha, M. R. Patel, and J. V. Patel, "Pet Waste Management by Chemical Recycling: A Review," *J. Polym. Environ.*, vol. 18, no. 1, pp. 8–25, Mar. 2010, doi: 10.1007/s10924-008-0106-7.

[18] M. Čolnik, Ž. Knez, and M. Škerget, "Sub- and supercritical water for chemical recycling of polyethylene terephthalate waste", *Chem. Eng. Sci.*, vol. 233, pp. 116389, Apr. 2021, doi: 10.1016/j.ces.2020.116389.

[19] L. Umdagas, R. Orozco, K. Heeley, W. Thom, and B. Al-Duri, "Advances in chemical recycling of polyethylene terephthalate (PET) via hydrolysis: A comprehensive review", *Polym. Degrad. Stab.*, vol. 234, pp. 111246, Apr. 2025, doi: 10.1016/j.polymdegradstab.2025.111246.

[20] C. Chaabani, E. Weiss-Hortala, and Y. Soudais, "Impact of Solvolysis Process on Both Depolymerization Kinetics of Nylon 6 and Recycling Carbon Fibres from Waste Composite," *Waste Biomass Valorization*, vol. 8, no. 8, pp. 2853–2865, Dec. 2017, doi: 10.1007/s12649-017-9901-5.

[21] V. Sinha, M. R. Patel, and J. V. Patel, "Pet waste management by chemical recycling: A review," *J. Polym. Environ.*, vol. 18, no. 1, pp. 8–25, 2010, doi: 10.1007/s10924-008-0106-7.

[22] M. Čolnik, Ž. Knez, and M. Škerget, "Sub- and supercritical water for chemical recycling of polyethylene terephthalate waste", *Chem. Eng. Sci.*, vol. 233, pp. 116389, Apr. 2021, doi: 10.1016/J.CES.2020.116389.

[23] C. Chaabani, E. Weiss-Hortala, and Y. Soudais, "Impact of Solvolysis Process on Both Depolymerization Kinetics of Nylon 6 and Recycling Carbon Fibres from Waste Composite," *Waste Biomass Valorization*, vol. 8, no. 8, pp. 2853–2865, 2017, doi: 10.1007/s12649-017-9901-5.

[24] T. Iwaya, M. Sasaki, and M. Goto, "Kinetic analysis for hydrothermal depolymerization of nylon 6," *Polym. Degrad. Stab.*, vol. 91, no. 9, pp. 1989–1995, Sep. 2006, doi: 10.1016/J.POLYMDEGRADSTAB.2006.02.009.

[25] W. Wang, L. Meng, and Y. Huang, "Hydrolytic degradation of monomer casting nylon in subcritical water", *Polym. Degrad. Stab.*, vol. 110, pp. 312–317, Dec. 2014, doi: 10.1016/J.POLYMDEGRADSTAB.2014.09.014.

[26] U. Češarek, D. Pahovnik, and E. Žagar, "Chemical Recycling of Aliphatic Polyamides by Microwave-Assisted Hydrolysis for Efficient Monomer Recovery", *ACS Sustain. Chem. Eng.*, vol. 8, pp. 16274–16282, 2020, doi: 10.1021/acssuschemeng.0c05706.

[27] S. R. Shukla, A. M. Harad, and D. Mahato, "Depolymerization of Nylon 6 Waste Fibres", *J. Appl. Polym. Sci.*, vol. 100, no. 1, 2005, doi: 10.1002/app.22775.

[28] S. Hocker, A. Rhudy, G. Ginsburg, and D. Kranbuehl, "Polyamide hydrolysis accelerated by small weak organic acids", *Polymer (Guildf)*, vol. 55, 2014, doi: 10.1016/j.polymer.2014.08.010.

[29] G. P. Karayannidis, A. P. Chatziavgoustis, and D. S. Achilias, "Poly(ethylene terephthalate) recycling and recovery of pure terephthalic acid by alkaline hydrolysis," *Adv. Polym. Technol.*, vol. 21, no. 4, pp. 250–259, Dec. 2002, doi: 10.1002/adv.10029.

[30] H. I. Khalaf, and O. A. Hasan, "Effect of quaternary ammonium salt as a phase transfer catalyst for the microwave depolymerization of polyethylene terephthalate waste bottles", *Chem. Eng. J.*, vol. 192, pp. 45–48, Jun. 2012, doi: 10.1016/j.cej.2012.03.081.

[31] N. R. Paliwal, and A. K. Mungray, "Ultrasound assisted alkaline hydrolysis of poly(ethylene terephthalate) in presence of phase transfer catalyst," *Polym. Degrad. Stab.*, vol. 98, no. 10, pp. 2094–2101, Oct. 2013, doi: 10.1016/j.polymdegradstab.2013.06.030.

[32] F. Forouzeshfar, H. Abedsoltan, M. R. Coleman, and J. G. Lawrence, "Alkaline hydrolysis of poly (ethylene terephthalate) using 1,8-diazabicyclo [5.4.0] undec-7-ene (DBU)", *Polym. Degrad. Stab.*, vol. 225, pp. 110814, Jul. 2024, doi: 10.1016/j.polymdegradstab.2024.110814.

[33] L. O. Dandy, G. Oliveux, J. Wood, M. J. Jenkins, and G. A. Leeke, "Accelerated degradation of Polyetheretherketone (PEEK) composite materials for recycling applications", *Polym. Degrad. Stab.*, vol. 112, pp. 52–62, Feb. 2015, doi: 10.1016/j.polymdegradstab.2014.12.012.

[34] L. Dandy, "Supercritical Fluids and Their Application to the Recycling of High-Performance Carbon Fibre Reinforced Composite Materials," University of Birmingham, Birmingham, 2015. Accessed: Nov. 12, 2024. [Online]. Available: http://etheses.bham.ac.uk/5896/

[35] S. M. Kurtz, "Synthesis and Processing of PEEK for Surgical Implants," in *PEEK Biomaterials Handbook*, Elsevier, 2019, pp. 11–25, doi: 10.1016/B978-0-12-812524-3.00002-8.

[36] L. O. Dandy, "Supercritical Fluids and Their Application to the Recycling of High-Performance Carbon Fibre Reinforced Composite Materials," *PhD Thesis*, no. April, 2015.

[37] J. J. Rubio Arias, and W. Thielemans, "Efficient Depolymerization of Glass Fibre Reinforced PET Composites," *Polymers (Basel)*, vol. 14, no. 23, pp. 5171, Nov. 2022, doi: 10.3390/polym14235171.

[38] R. Scaffaro, A. Di Bartolo, and N. Tz. Dintcheva, "Matrix and Filler Recycling of Carbon and Glass Fibre-Reinforced Polymer Composites: A Review," *Polymers (Basel)*, vol. 13, no. 21, pp. 3817, Nov. 2021, doi: 10.3390/polym13213817.

[39] B. -. K. Kim, G. -. C. Hwang, S. -. Y. Bae, S. -. C. Yi, and H. Kumazawa, "Depolymerization of polyethyleneterephthalate in supercritical methanol," *J. Appl. Polym. Sci.*, vol. 81, no. 9, pp. 2102–2108, Aug. 2001, doi: 10.1002/app.1645.

[40] National Centre for Biotechnology Information, "Dimethyl terephthalate," PubChem. Accessed: Sep. 12, 2025. [Online]. Available: https://pubchem.ncbi.nlm.nih.gov/compound/Dimethyl-terephthalate

[41] H. Liu, *et al.*, Lewis Acidic Ionic Liquid [Bmim]FeCl4 as a High Efficient Catalyst for Methanolysis of Poly (lactic acid), *Catal. Lett.*, vol. 147, no. 9, pp. 2298–2305, Sep. 2017, doi: 10.1007/s10562-017-2138-x.

[42] A. Kamimura, *et al.*, Direct conversion of polyamides to ω-hydroxyalkanoic acid derivatives by using supercritical MeOH, *Green Chem.*, vol. 13, no. 8, pp. 2055, 2011, doi: 10.1039/c1gc15172j.

[43] T. Kaweetirawatt, T. Yamaguchi, S. Hayashiyama, M. Sumimoto, A. Kamimura, and K. Hori, "Nylon 6 depolymerization in supercritical alcohols studied by the QM/MC/FEP method," *RSC Adv.*, vol. 2, no. 22, pp. 8402, 2012, doi: 10.1039/c2ra20835k.

[44] W. T. Lang, S. A. Mehta, M. M. Thomas, D. Openshaw, E. Westgate, and G. Bagnato, "Chemical recycling of polyethylene terephthalate, an industrial and sustainable opportunity for Northwest of England," *J. Environ. Chem. Eng.*, vol. 11, no. 5, pp. 110585, Oct. 2023, doi: 10.1016/j.jece.2023.110585.

[45] J. Sutton, G. Grause, A. Al Rida Hmayed, S. T. G. Street, A. P. Dove, and J. Wood, "Organocatalytic glycolysis of polyethylene terephthalate and product separation by membrane filtration", *Chem. Eng. J.*, vol. 512, pp. 162400, May. 2025, doi: 10.1016/j.cej.2025.162400.

[46] H. Ando, M. Oshima, Y. Nakayama, and A. Nakayama, "Polyethylene glycol-solvolyzed poly-(l)-lactic acids and their stereocomplexes with poly-(d)-lactic acid," *Polym. Degrad. Stab.*, vol. 98, no. 5, pp. 958–962, May. 2013, doi: 10.1016/j.polymdegradstab.2013.02.016.

[47] R. J. Tapper, M. L. Longana, I. Hamerton, and K. D. Potter, "A closed-loop recycling process for discontinuous carbon fibre polyamide 6 composites," *Compos. Part B*, vol. 179, no. August, pp. 107418, 2019, doi: 10.1016/j.compositesb.2019.107418.

[48] F. Knappich, M. Klotz, M. Schlummer, J. Wölling, and A. Mäurer, "Recycling process for carbon fibre reinforced plastics with polyamide 6, polyurethane and epoxy matrix by gentle solvent treatment", *Waste Manage.*, vol. 85, pp. 73–81, 2019, doi: 10.1016/j.wasman.2018.12.016.

[49] M. Zhang, *et al.*, Targeted valorization of waste polycarbonate into bisphenol A dimethyl ether, *Chem. Eng. J.*, vol. 500, pp. 156914, Nov. 2024, doi: 10.1016/j.cej.2024.156914.

[50] F. D'Anna, G. Raia, G. Di Cara, P. Cancemi, and S. Marullo, "Task-specific ionic liquids and ultrasound irradiation: a successful strategy to drive the alcoholysis of polycarbonate," *RSC Sustainability*, vol. 3, no. 1, pp. 580–591, 2025, doi: 10.1039/D4SU00301B.

[51] R. J. Tapper, M. L. Longana, H. Yu, I. Hamerton, and K. D. Potter, "Development of a closed-loop recycling process for discontinuous carbon fibre polypropylene composites", *Compos. B Eng.*, vol. 146, pp. 222–231, Aug. 2018, doi: 10.1016/J.COMPOSITESB.2018.03.048.

[52] J. G. Poulakis, P. C. Varelidis, and C. D. Papaspyrides, "Recycling of Polypropylene- Based Composites," *Adv. Polym. Technol.*, vol. 16, no. 4, pp. 251–336, 1997.

[53] D. S. Cousins, Y. Suzuki, R. E. Murray, J. R. Samaniuk, and A. P. Stebner, "Recycling glass fibre thermoplastic composites from wind turbine blades", *J. Clean. Prod.*, vol. 209, pp. 1252–1263, Feb. 2019, doi: 10.1016/J.JCLEPRO.2018.10.286.

[54] I. Y. Evchuk, R. I. Musii, R. G. Makitra, and R. E. Pristanskii, "Solubility of Polymethyl Methacrylate in Organic Solvents," *Russ. J. Appl. Chem.*, vol. 78, no. 10, pp. 1576–1580, 2005.

[55] C. Tschentscher, M. Gebhardt, S. Chakraborty, and D. Meiners, "Recycling of Elium CFRPs for high temperature dissolution: A study with different solvents," in *Materialtechnik Symposium 2021*, Clausthal, 2021, pp. 1–12.

[56] J. Zhang, V. S. Chevali, H. Wang, and C. H. Wang, "Current status of carbon fibre and carbon fibre composites recycling," *Compos. B Eng.*, vol. 193, no. December 2019, pp. 108053, 2020, doi: 10.1016/j.compositesb.2020.108053.

[57] L. C. Bank, and A. Yazdanbakhsh, "Reuse of glass thermoset FRP composites in the construction industry – A growing opportunity," in *Proceedings of the 7th International Conference on FRP Composites in Civil Engineering, CICE 2014*, pp. 1–6, 2014.

[58] G. Oliveux, J. L. Bailleul, E. L. G. La Salle, N. Lefèvre, and G. Biotteau, "Recycling of glass fibre reinforced composites using subcritical hydrolysis: Reaction mechanisms and kinetics, influence of the chemical structure of the resin," *Polym. Degrad. Stab.*, vol. 98, no. 3, pp. 785–800, 2013, doi: 10.1016/j.polymdegradstab.2012.12.010.

[59] H. Sato, *et al.*, Sixteen Thousand Evaluated Experimental Thermodynamic Property Data for Water and Steam, *J. Phys. Chem. Ref. Data*, vol. 20, pp. 1023, 1991.

[60] Y. N. Kim, *et al.*, Application of supercritical water for green recycling of epoxy-based carbon fibre reinforced plastic, *Compos. Sci. Technol.*, vol. 173, no. July 2018, pp. 66–72, 2019, doi: 10.1016/j.compscitech.2019.01.026.

[61] R. Piñero-Hernanz, *et al.*, Chemical recycling of carbon fibre reinforced composites in nearcritical and supercritical water, *Compos. Part A Appl. Sci. Manuf.*, vol. 39, no. 3, pp. 454–461, Mar. 2008, doi: 10.1016/J.COMPOSITESA.2008.01.001.

[62] L. Henry, A. Schneller, J. Doerfler, W. M. Mueller, C. Aymonier, and S. Horn, "Semi-continuous flow recycling method for carbon fibre reinforced thermoset polymers by near- and supercritical solvolysis", *Polym. Degrad. Stab.*, vol. 133, pp. 264–274, Nov. 2016, doi: 10.1016/J.POLYMDEGRADSTAB.2016.09.002.

[63] Y. Bai, Z. Wang, and L. Feng, "Chemical recycling of carbon fibres reinforced epoxy resin composites in oxygen in supercritical water," *Mater. Des.*, vol. 31, no. 2, pp. 999–1002, 2010, doi: 10.1016/j.matdes.2009.07.057.

[64] L. Yuyan, S. Guohua, and M. Linghui, "Recycling of carbon fibre reinforced composites using water in subcritical conditions," *Mater. Sci. Eng. A*, vol. 520, no. 1–2, pp. 179–183, 2009, doi: 10.1016/j.msea.2009.05.030.

[65] G. Oliveux, J.-L. Bailleul, and E. L. G. La Salle, "Chemical recycling of glass fibre reinforced composites using subcritical water," *Compos. Part A Appl. Sci. Manuf.*, vol. 43, no. 11, pp. 1809–1818, Nov. 2012, doi: 10.1016/j.compositesa.2012.06.008.

[66] G. Oliveux, L. O. Dandy, and G. A. Leeke, "Degradation of a model epoxy resin by solvolysis routes", *Polym. Degrad. Stab.*, vol. 118, pp. 96–103, Aug. 2015, doi: 10.1016/j.polymdegradstab.2015.04.016.

[67] M. J. Keith, L. A. Román-Ramírez, G. Leeke, and A. Ingram, "Recycling a carbon fibre reinforced polymer with a supercritical acetone/water solvent mixture: Comprehensive analysis of reaction kinetics", *Polym. Degrad. Stab.*, vol. 161, pp. 225–234, Mar. 2019, doi: 10.1016/j.polymdegradstab.2019.01.015.

[68] L. Henry, A. Schneller, J. Doerfler, W. M. Mueller, C. Aymonier, and S. Horn, "Semi-continuous flow recycling method for carbon fibre reinforced thermoset polymers by near- and supercritical solvolysis", *Polym. Degrad. Stab.*, vol. 133, pp. 264–274, Nov. 2016, doi: 10.1016/j.polymdegradstab.2016.09.002.

[69] W. Dang, M. Kubouchi, S. Yamamoto, H. Sembokuya, and K. Tsuda, "An approach to chemical recycling of epoxy resin cured with amine using nitric acid," *Polymer (Guildf)*, vol. 43, no. 10, pp. 2953–2958, May. 2002, doi: 10.1016/S0032-3861(02)00100-3.

[70] Y. Liu, L. Meng, Y. Huang, and J. Du, "Recycling of carbon/epoxy composites," *J. Appl. Polym. Sci.*, vol. 94, no. 5, pp. 1912–1916, Dec. 2004, doi: 10.1002/app.20990.

[71] S.-H. Lee, H.-O. Choi, J.-S. Kim, C.-K. Lee, Y.-K. Kim, and C.-S. Ju, "Circulating flow reactor for recycling of carbon fibre from carbon fibre reinforced epoxy composite," *Korean J. Chem. Eng.*, vol. 28, no. 2, pp. 449–454, Feb. 2011, doi: 10.1007/s11814-010-0394-1.

[72] P. Feraboli, H. Kawakami, B. Wade, F. Gasco, L. DeOto, and A. Masini, "Recyclability and reutilization of carbon fibre fabric/epoxy composites," *J. Compos. Mater*, vol. 46, no. 12, pp. 1459–1473, Jun. 2012, doi: 10.1177/0021998311420604.

[73] L. Yuyan, S. Guohua, and M. Linghui, "Recycling of carbon fibre reinforced composites using water in subcritical conditions," *Mater. Sci. Eng. A*, vol. 520, no. 1–2, pp. 179–183, Sep. 2009, doi: 10.1016/j.msea.2009.05.030.

[74] A. Yamaguchi, *et al.*, Recyclable carbon fibre-reinforced plastics containing degradable acetal linkages: Synthesis, properties, and chemical recycling, *J. Polym. Sci. A Polym. Chem.*, vol. 53, no. 8, pp. 1052–1059, Apr. 2015, doi: 10.1002/pola.27575.

[75] N. Zhang, *et al.*, Amphiphilic catalyst for decomposition of unsaturated polyester resins to valuable chemicals with 100% atom utilization efficiency, *J. Clean. Prod.*, vol. 296, pp. 126492, May. 2021, doi: 10.1016/j.jclepro.2021.126492.

[76] Y. Liu, J. Liu, Z. Jiang, and T. Tang, "Chemical recycling of carbon fibre reinforced epoxy resin composites in subcritical water: Synergistic effect of phenol and KOH on the decomposition efficiency," *Polym. Degrad. Stab.*, vol. 97, no. 3, pp. 214–220, Mar. 2012, doi: 10.1016/J.POLYMDEGRADSTAB.2011.12.028.

[77] M. J. Keith, G. A. Leeke, P. Khan, and A. Ingram, "Catalytic degradation of a carbon fibre reinforced polymer for recycling applications", *Polym. Degrad. Stab.*, vol. 166, pp. 188–201, Aug. 2019, doi: 10.1016/j.polymdegradstab.2019.05.020.

[78] T. Deng, *et al.*, Cleavage of C-N bonds in carbon fibre/epoxy resin composites, *Green Chem.*, vol. 17, no. 4, pp. 2141–2145, 2015, doi: 10.1039/c4gc02512a.

[79] T. Liu, *et al.*, Mild chemical recycling of aerospace fibre/epoxy composite wastes and utilization of the decomposed resin, *Polym. Degrad. Stab.*, vol. 139, pp. 20–27, May. 2017, doi: 10.1016/j.polymdegradstab.2017.03.017.

[80] C. Hao, *et al.*, Mild chemical recycling of waste wind turbine blade for direct reuse in production of thermoplastic composites with enhanced performance, *Resour. Conserv. Recycl.*, vol. 215, pp. 108159, Apr. 2025, doi: 10.1016/j.resconrec.2025.108159.

[81] L. Ye, K. Wang, H. Feng, and Y. Wang, "Recycling of Carbon Fibre-reinforced Epoxy Resin-based Composites Using a Benzyl Alcohol/Alkaline System," *Fibres Polym.*, vol. 22, no. 3, pp. 811–818, Mar. 2021, doi: 10.1007/s12221-021-0266-9.

[82] R. Piñero-Hernanz, *et al.*, Chemical recycling of carbon fibre composites using alcohols under subcritical and supercritical conditions, *J. Supercrit. Fluids*, vol. 46, no. 1, pp. 83–92, Aug. 2008, doi: 10.1016/j.supflu.2008.02.008.

[83] I. Okajima, M. Hiramatsu, Y. Shimamura, T. Awaya, and T. Sako, "Chemical recycling of carbon fibre reinforced plastic using supercritical methanol", *J. Supercrit. Fluids*, vol. 91, pp. 68–76, Jul. 2014, doi: 10.1016/j.supflu.2014.04.011.

[84] I. Okajima, K. Watanabe, S. Haramiishi, M. Nakamura, Y. Shimamura, and T. Sako, "Recycling of carbon fibre reinforced plastic containing amine-cured epoxy resin using supercritical and subcritical fluids", *J. Supercrit. Fluids*, vol. 119, pp. 44–51, Jan. 2017, doi: 10.1016/j.supflu.2016.08.015.

[85]   G. Jiang, S. Pickering, E. Lester, T. Turner, K. Wong, and N. Warrior, "Characterisation of carbon fibres recycled from carbon fibre/epoxy resin composites using supercritical n-propanol," *Compos. Sci. Technol.*, vol. 69, no. 2, pp. 192–198, Feb. 2009, doi: 10.1016/j.compscitech.2008.10.007.

[86]   H. Yan, C. Lu, D. Jing, and X. Hou, "Chemical degradation of amine-cured DGEBA epoxy resin in supercritical 1-propanol for recycling carbon fibre from composites," *Chin. J. Polym. Sci.*, vol. 32, no. 11, pp. 1550–1563, Nov. 2014, doi: 10.1007/s10118-014-1519-5.

[87]   Q. Zhao, L. An, C. Li, L. Zhang, J. Jiang, and Y. Li, "Environment-friendly recycling of CFRP composites via gentle solvent system at atmospheric pressure", *Compos. Sci. Technol.*, vol. 224, pp. 109461, Jun. 2022, doi: 10.1016/j.compscitech.2022.109461.

[88]   X. Zhao, X.-L. Wang, F. Tian, W.-L. An, S. Xu, and Y.-Z. Wang, "A fast and mild closed-loop recycling of anhydride-cured epoxy through microwave-assisted catalytic degradation by trifunctional amine and subsequent reuse without separation," *Green Chem.*, vol. 21, no. 9, pp. 2487–2493, 2019, doi: 10.1039/C9GC00685K.

[89]   Z. Fehér, *et al.*, Optimisation of PET glycolysis by applying recyclable heterogeneous organocatalysts, *Green Chem.*, vol. 24, no. 21, pp. 8447–8459, 2022, doi: 10.1039/D2GC02860C.

[90]   P. Yang, Q. Zhou, X.-Y. Li, -K.-K. Yang, and Y.-Z. Wang, "Chemical recycling of fibre-reinforced epoxy resin using a polyethylene glycol/NaOH system," *J. Reinf. Plast. Compos.*, vol. 33, no. 22, pp. 2106–2114, Nov. 2014, doi: 10.1177/0731684414555745.

[91]   J. Jiang, *et al.*, On the successful chemical recycling of carbon fibre/epoxy resin composites under the mild condition, *Compos. Sci. Technol.*, vol. 151, pp. 243–251, Oct. 2017, doi: 10.1016/j.compscitech.2017.08.007.

[92]   P. Yang, Q. Zhou, -X.-X. Yuan, J. M. N. Van kasteren, and Y.-Z. Wang, "Highly efficient solvolysis of epoxy resin using poly(ethylene glycol)/NaOH systems," *Polym. Degrad. Stab.*, vol. 97, no. 7, pp. 1101–1106, Jul. 2012, doi: 10.1016/j.polymdegradstab.2012.04.007.

[93]   P. Xu, J. Li, and J. Ding, "Chemical recycling of carbon fibre/epoxy composites in a mixed solution of peroxide hydrogen and N,N-dimethylformamide", *Compos. Sci. Technol.*, vol. 82, pp. 54–59, Jun. 2013, doi: 10.1016/J.COMPSCITECH.2013.04.002.

[94]   J. Li, P.-L. Xu, Y.-K. Zhu, J.-P. Ding, L.-X. Xue, and Y.-Z. Wang, "A promising strategy for chemical recycling of carbon fibre/thermoset composites: self-accelerating decomposition in a mild oxidative system," *Green Chem.*, vol. 14, no. 12, pp. 3260, 2012, doi: 10.1039/c2gc36294e.

[95]   M. Das, R. Chacko, and S. Varughese, "An Efficient Method of Recycling of CFRP Waste Using Peracetic Acid," *ACS Sustain. Chem. Eng.*, vol. 6, no. 2, pp. 1564–1571, Feb. 2018, doi: 10.1021/acssuschemeng.7b01456.

[96]   Y. Wang, *et al.*, Chemical recycling of unsaturated polyester resin and its composites via selective cleavage of the ester bond, *Green Chem.*, vol. 17, no. 9, pp. 4527–4532, 2015, doi: 10.1039/C5GC01048A.

[97]   E. L. Smith, A. P. Abbott, and K. S. Ryder, "Deep Eutectic Solvents (DESs) and Their Applications," *Chem. Rev.*, vol. 114, no. 21, pp. 11060–11082, Nov. 2014, doi: 10.1021/cr300162p.

[98]   C.-W. Liu, W.-J. Hong, B.-T. Yang, C.-W. Lin, L.-C. Wang, and -C.-C. Chen, "Switchable deep eutectic solvents as efficient and sustainable recycling media for carbon fibre reinforced polymer composite waste", *J. Clean. Prod.*, vol. 378, pp. 134334, Dec. 2022, doi: 10.1016/j.jclepro.2022.134334.

[99]   O. Levenspiel, *Chemical Reaction Engineering*, New Jersey: John Wiley & Sons, 1999.

[100] I. Okajima, K. Watanabe, S. Haramiishi, M. Nakamura, Y. Shimamura, and T. Sako, "Recycling of carbon fibre reinforced plastic containing amine-cured epoxy resin using supercritical and subcritical fluids", *J. Supercrit. Fluids*, vol. 119, pp. 44–51, 2017, doi: 10.1016/j.supflu.2016.08.015.

[101] W. Dang, M. Kubouchi, H. Sembokuya, and K. Tsuda, "Chemical recycling of glass fibre reinforced epoxy resin cured with amine using nitric acid," *Polymer (Guildf)*, vol. 46, no. 6, pp. 1905–1912, Feb. 2005, doi: 10.1016/J.POLYMER.2004.12.035.

[102] M. L. Longana, N. Ong, H. Yu, and K. D. Potter, "Multiple closed loop recycling of carbon fibre composites with the HiPerDiF (High Performance Discontinuous Fibre) method", *Compos. Struct.*, vol. 153, pp. 271–277, Oct. 2016, doi: 10.1016/j.compstruct.2016.06.018.

[103] K. I. Ismail, T. C. Yap, and R. Ahmed, "3D-Printed Fibre-Reinforced Polymer Composites by Fused Deposition Modelling (FDM): Fibre Length and Fibre Implementation Techniques," *Polymers (Basel)*, vol. 14, no. 21, pp. 4659, Nov. 2022, doi: 10.3390/polym14214659.

[104] H. Yan, C. Lu, D. Jing, C. Chang, N. Liu, and X. Hou, "Recycling of carbon fibres in epoxy resin composites using supercritical 1-propanol," *New Carbon Mater.*, vol. 31, no. 1, pp. 46–54, Feb. 2016, doi: 10.1016/S1872-5805(16)60004-5.

[105] W. Liu, H. Huang, H. Cheng, and Z. Liu, "CFRP Reclamation and Remanufacturing Based on a Closed-loop Recycling Process for Carbon Fibres Using Supercritical N-butanol," *Fibres Polym.*, vol. 21, no. 3, pp. 604–618, Mar. 2020, doi: 10.1007/s12221-020-9575-7.

[106] R. J. Tapper, M. L. Longana, I. Hamerton, and K. D. Potter, "A closed-loop recycling process for discontinuous carbon fibre polyamide 6 composites", *Compos. B Eng.*, vol. 179, pp. 107418, Dec. 2019, doi: 10.1016/j.compositesb.2019.107418.

[107] Y. Wang, *et al.*, Chemical Recycling of Carbon Fibre Reinforced Epoxy Resin Composites via Selective Cleavage of the Carbon–Nitrogen Bond, *ACS Sustain. Chem. Eng.*, vol. 3, no. 12, pp. 3332–3337, Dec. 2015, doi: 10.1021/acssuschemeng.5b00949.

[108] M. F. Muhammad Faisal, A. Hassan, K. W. Gan, M. N. Roslan, and A. H. Abdul Rashid, "Effects of Sulphuric Acid Concentrations during Solvolysis Process of Carbon Fibre Reinforced Epoxy Composite," *Sains Malays*, vol. 49, no. 09, pp. 2073–2081, Sep. 2020, doi: 10.17576/jsm-2020-4909-05.

[109] H. Feng, *et al.*, Facile preparation, closed-loop recycling of multifunctional carbon fibre reinforced polymer composites, *Compos. B Eng.*, vol. 257, pp. 110677, May. 2023, doi: 10.1016/j. compositesb.2023.110677.

[110] Y. Wang, *et al.*, Chemical Recycling of Carbon Fibre Reinforced Epoxy Resin Composites via Selective Cleavage of the Carbon–Nitrogen Bond, *ACS Sustain. Chem. Eng.*, vol. 3, no. 12, pp. 3332–3337, Dec. 2015, doi: 10.1021/acssuschemeng.5b00949.

[111] Q. Zhao, *et al.*, Controlling degradation and recycling of carbon fibre reinforced bismaleimide resin composites via selective cleavage of imide bonds, *Compos. B Eng.*, vol. 231, pp. 109595, Feb. 2022, doi: 10.1016/j.compositesb.2021.109595.

[112] Q. Qiu, and M. Kumosa, "Corrosion of E-glass fibres in acidic environments," *Compos. Sci. Technol.*, vol. 57, no. 5, pp. 497–507, 1997, doi: 10.1016/S0266-3538(96)00158-3.

[113] H. U. Sokoli, J. Beauson, M. E. Simonsen, A. Fraisse, P. Brøndsted, and E. G. Søgaard, "Optimized process for recovery of glass- and carbon fibres with retained mechanical properties by means of near- and supercritical fluids", *J. Supercrit. Fluids*, vol. 124, pp. 80–89, Jun. 2017, doi: 10.1016/j. supflu.2017.01.013.

[114] C. C. Kao, O. R. Ghita, K. R. Hallam, P. J. Heard, and K. E. Evans, "Mechanical studies of single glass fibres recycled from hydrolysis process using sub-critical water," *Compos. Part A Appl. Sci. Manuf.*, vol. 43, no. 3, pp. 398–406, Mar. 2012, doi: 10.1016/j.compositesa.2011.11.011.

[115] A. Kamimura, *et al.*, DMAP as an Effective Catalyst To Accelerate the Solubilization of Waste Fibre-Reinforced Plastics, *ChemSusChem*, vol. 1, no. 10, pp. 845–850, Oct. 2008, doi: 10.1002/ cssc.200800151.

[116] J.-Y. Lee, and K.-J. Kim, "MEG Effects on Hydrolysis of Polyamide 66/Glass Fibre Composites and Mechanical Property Changes," *Molecules*, vol. 24, no. 4, pp. 755, Feb. 2019, doi: 10.3390/ molecules24040755.

[117] P. R. Souza, *et al.*, Sub- and supercritical D-limonene technology as a green process to recover glass fibres from glass fibre-reinforced polyester composites, *J. Clean. Prod.*, vol. 254, pp. 119984, May. 2020, doi: 10.1016/j.jclepro.2020.119984.

[118] V. Schenk, K. Labastie, M. Destarac, P. Olivier, and M. Guerre, "Vitrimer composites: current status and future challenges," *Mater. Adv.*, vol. 3, no. 22, pp. 8012–8029, 2022, doi: 10.1039/D2MA00654E.

[119] J. H. Jeon, *et al.*, Assessment of recycling and repair methods for discontinuous glass fibre reinforced vitrimer composites with reclaimed parts, *Manuf. Lett.*, vol. 41, pp. 1659–1668, Oct. 2024, doi: 10.1016/j.mfglet.2024.09.193.

[120] J. Zhao, P. Liu, J. Yue, H. Huan, G. Bi, and L. Zhang, "Recycling glass fibres from thermoset epoxy composites by in situ oxonium-type polyionic liquid formation and naphthalene-containing superplasticizer synthesis with the degradation solution of the epoxy resin", *Compos. B Eng.*, vol. 254, pp. 110435, Apr. 2023, doi: 10.1016/j.compositesb.2022.110435.

[121] L. Yuyan, M. Linghui, H. Yudong, and L. Lixun, "Method of Recovering the Fibrous Fraction of Glass/ Epoxy Composites," *J. Reinf. Plast. Compos.*, vol. 25, no. 14, pp. 1525–1533, Sep. 2006, doi: 10.1177/ 0731684406066748.

[122] S. Harisankar, and P. Biller, "Kinetics and recovery of bisphenol-A from fibre-reinforced polycarbonate using subcritical water", *React. Chem. Eng.*, 2025, doi: 10.1039/D5RE00291E.

[123] S. T. Bashir, L. Yang, J. J. Liggat, and J. L. Thomason, "Kinetics of dissolution of glass fibre in hot alkaline solution," *J. Mater. Sci.*, vol. 53, no. 3, pp. 1710–1722, Feb. 2018, doi: 10.1007/s10853-017- 1627-z.

[124] The Essential Chemical Industry, "Paints." Accessed: Oct. 10, 2025. [Online]. Available: https://www. essentialchemicalindustry.org/materials-and-applications/paints.html

[125] Sigma Aldrich, "Paints & Coatings Solvents Air Monitoring Applications." Accessed: Oct. 10, 2025. [Online]. Available: https://www.sigmaaldrich.com/GB/en/technical-documents/technical-article/en vironmental-testing-and-industrial-hygiene/air-testing/paints-and-coatings-solvents-air- monitoring?srsltid=AfmBOoq7rBoEBlkTn2FOvDMYSkQlYTO675qivYmcTY6sRUXaYAhP-GjF

[126] K. Hunger, and W. Herbst, "Pigments, Organic," in *Ullmann's Encyclopedia of Industrial Chemistry*, Wiley, 2000, doi: 10.1002/14356007.a20_371.

[127] M. Liska, A. Wilson, and J. Bensted, "Special Cements," in *Lea's Chemistry of Cement and Concrete*, Elsevier, 2019, pp. 585–640, doi: 10.1016/B978-0-08-100773-0.00013-7.

[128] Z. Dohnalová, L. Svoboda, and P. Sulcová, "Characterization of kaolin dispersion using acoustic and electroacoustic spectroscopy," *J. Mini. Metal. Sect. B Metall.*, vol. 44, no. 1, pp. 63–72, 2008, doi: 10.2298/JMMB0801063D.

[129] Z. YAN, Z. WANG, X. WANG, H. LIU, and J. QIU, "Kinetic model for calcium sulfate decomposition at high temperature," *Trans. Nonferrous Metal. Soc. China*, vol. 25, no. 10, pp. 3490–3497, Oct. 2015, doi: 10.1016/S1003-6326(15)63986-3.

[130] K. Ishikawa, "Bioactive Ceramics: Cements," in *Comprehensive Biomaterials*, Elsevier, 2011, pp. 267–283, doi: 10.1016/B978-0-08-055294-1.00029-5.

[131] S. Gharde, and B. Kandasubramanian, "Mechanothermal and chemical recycling methodologies for the Fibre Reinforced Plastic (FRP)", *Environ. Technol. Innov.*, vol. 14, pp. 100311, May. 2019, doi: 10.1016/j.eti.2019.01.005.

[132] J. Sharma, S. Shukla, G. V. Ramana, and B. K. Behera, "Advances in carbon and glass fibre recycling: optimal composite recycling and sustainable solutions for composite waste," *J. Mater. Cycles. Waste Manag.*, vol. 27, no. 5, pp. 3166–3195, Sep. 2025, doi: 10.1007/s10163-025-02342-0.

[133] K. Yu, "An environmental and economic study on the chemical recycling of plastic composites using an engineering constitutive model", *J. Environ. Manage.*, vol. 381, pp. 125271, May. 2025, doi: 10.1016/ j.jenvman.2025.125271.

[134] N. Poranek, *et al.*, Comparative LCA Analysis of Selected Recycling Methods for Carbon Fibres and Socio-Economic Analysis, *Materials*, vol. 18, no. 11, pp. 2660, Jun. 2025, doi: 10.3390/ma18112660.

[135] A. K. Kamali, J. Isayev, B. Laratte, and G. Sonnemann, "Harmonizing life cycle assessment studies of emerging technologies: The case of virgin and recycled carbon fibres", *Resour. Conserv. Recycl.*, vol. 220, pp. 108323, Jun. 2025, doi: 10.1016/j.resconrec.2025.108323.

[136] C. Vogiantzi, and K. Tserpes, "A Comparative Environmental and Economic Analysis of Carbon Fibre-Reinforced Polymer Recycling Processes Using Life Cycle Assessment and Life Cycle Costing," *J. Compos. Sci.*, vol. 9, no. 1, pp. 39, Jan. 2025, doi: 10.3390/jcs9010039.

[137] Y. F. Khalil, "Comparative environmental and human health evaluations of thermolysis and solvolysis recycling technologies of carbon fibre reinforced polymer waste", *Waste Manage.*, vol. 76, pp. 767–778, Jun. 2018, doi: 10.1016/j.wasman.2018.03.026.

[138] S. Karuppannan Gopalraj, and T. Kärki, "A review on the recycling of waste carbon fibre/glass fibre-reinforced composites: fibre recovery, properties and life-cycle analysis," *SN Appl. Sci.*, vol. 2, no. 3, pp. 433, Mar. 2020, doi: 10.1007/s42452-020-2195-4.

[139] B. Pillain, *et al.*, Positioning supercritical solvolysis among innovative recycling and current waste management scenarios for carbon fibre reinforced plastics thanks to comparative life cycle assessment, *J. Supercrit. Fluids*, vol. 154, pp. 104607, Dec. 2019, doi: 10.1016/j.supflu.2019.104607.

[140] E. Pakdel, S. Kashi, R. Varley, and X. Wang, "Recent progress in recycling carbon fibre reinforced composites and dry carbon fibre wastes", *Resour. Conserv. Recycl.*, vol. 166, pp. 105340, Mar. 2021, doi: 10.1016/j.resconrec.2020.105340.

[141] M. Prinçaud, C. Aymonier, A. Loppinet-Serani, N. Perry, and G. Sonnemann, "Environmental Feasibility of the Recycling of Carbon Fibres from CFRPs by Solvolysis Using Supercritical Water," *ACS Sustain. Chem. Eng.*, vol. 2, no. 6, pp. 1498–1502, Jun. 2014, doi: 10.1021/sc500174m.

[142] K.-R. Chatzipanagiotou, *et al.*, Life Cycle Assessment of Composites Additive Manufacturing Using Recycled Materials, *Sustainability*, vol. 15, no. 17, pp. 12843, Aug. 2023, doi: 10.3390/su151712843.

[143] K. Kooduvalli, J. Unser, S. Ozcan, and U. K. Vaidya, "Embodied Energy in Pyrolysis and Solvolysis Approaches to Recycling for Carbon Fibre-Epoxy Reinforced Composite Waste Streams," *Recycling*, vol. 7, no. 1, pp. 6, Feb. 2022, doi: 10.3390/recycling7010006.

[144] K. Kawajiri, and M. Kobayashi, "Cradle-to-Gate life cycle assessment of recycling processes for carbon fibres: A case study of ex-ante life cycle assessment for commercially feasible pyrolysis and solvolysis approaches", *J. Clean. Prod.*, vol. 378, pp. 134581, Dec. 2022, doi: 10.1016/j.jclepro.2022.134581.

[145] E. Urruzola, *et al.*, Eco-efficiency assessment and benchmarking of recycled carbon fibre, *Cleaner Mater.*, vol. 17, pp. 100333, Sep. 2025, doi: 10.1016/j.clema.2025.100333.

[146] Z. Alavi, K. Khalilpour, N. Florin, A. Hadigheh, and A. Hoadley, "End-of-life wind turbine blade management across energy transition: A life cycle analysis", *Resour. Conserv. Recycl.*, vol. 213, pp. 108008, Feb. 2025, doi: 10.1016/j.resconrec.2024.108008.

[147] L. Merlo-Camuñas, E. Urruzola, E. De la guerra, M. Azcona, and D. Iribarren, "Environmental life-cycle performance of alternative pieces for trains based on the use of recycled carbon fibre", *J. Clean. Prod.*, vol. 452, pp. 142157, May. 2024, doi: 10.1016/j.jclepro.2024.142157.

[148] M. Wu, J. Sadhukhan, R. Murphy, U. Bharadwaj, and X. Cui, "A novel life cycle assessment and life cycle costing framework for carbon fibre-reinforced composite materials in the aviation industry," *Int. J. Life Cycle Assess.*, vol. 28, no. 5, pp. 566–589, May. 2023, doi: 10.1007/s11367-023-02164-y.

[149] S. Sobek, *et al.*, A life cycle assessment of the laboratory—scale oxidative liquefaction as the chemical recycling method of the end-of-life wind turbine blades, *J. Environ. Manage.*, vol. 361, pp. 121241, Jun. 2024, doi: 10.1016/j.jenvman.2024.121241.

[150] Y. Wei, and S. A. Hadigheh, "Cost benefit and life cycle analysis of CFRP and GFRP waste treatment methods", *Constr. Build. Mater.*, vol. 348, pp. 128654, Sep. 2022, doi: 10.1016/j.conbuildmat.2022.128654.

[151] A. D. La Rosa, S. Greco, C. Tosto, and G. Cicala, "LCA and LCC of a chemical recycling process of waste CF-thermoset composites for the production of novel CF-thermoplastic composites. Open loop and closed loop scenarios", *J. Clean. Prod.*, vol. 304, pp. 127158, Jul. 2021, doi: 10.1016/j.jclepro.2021.127158.

[152] A. D. La Rosa, D. R. Banatao, S. J. Pastine, A. Latteri, and G. Cicala, "Recycling treatment of carbon fibre/epoxy composites: Materials recovery and characterization and environmental impacts through life cycle assessment", *Compos. B Eng.*, vol. 104, pp. 17–25, Nov. 2016, doi: 10.1016/j. compositesb.2016.08.015.

[153] R. J. Tapper, M. L. Longana, A. Norton, K. D. Potter, and I. Hamerton, "An evaluation of life cycle assessment and its application to the closed-loop recycling of carbon fibre reinforced polymers", *Compos. B Eng.*, vol. 184, pp. 107665, Mar. 2020, doi: 10.1016/j.compositesb.2019.107665.

[154] P. A. Vo Dong, C. Azzaro-Pantel, and A.-L. Cadene, "Economic and environmental assessment of recovery and disposal pathways for CFRP waste management", *Resour. Conserv. Recycl.*, vol. 133, pp. 63–75, Jun. 2018, doi: 10.1016/j.resconrec.2018.01.024.

[155] A. McGregor, "UK plastic packaging tax data shows environmental and economic impact," Out-law News. Accessed: Oct. 10, 2025. [Online]. Available: https://www.pinsentmasons.com/out-law/news/ plastic-packaging-tax-data-environmental-economic-impact

[156] Panasonic Corporation, "Accelerating Recycling-Oriented Manufacturing with Plant-Derived Cellulose Fibre," *Panasonic Newsroom*, Oct. 2020. Accessed: Oct. 10, 2025. [Online]. Available: https://news.panasonic.com/global/stories/899

[157] Ledger Insights, "Fujitsu, chemicals firm Teijin use blockchain to trace recycled carbon fibre." Accessed: Oct. 10, 2025. [Online]. Available: https://www.ledgerinsights.com/fujitsu-chemicals-firm-teijin-use-blockchain-to-trace-recycled-carbon-fibre/

[158] Adherent Technologies Inc., "Recycling Technologies." Accessed: Nov. 19, 2024. [Online]. Available: https://www.adherent-tech.com/recycling_technologies/

[159] Uplift360, "Resin Removal – ChemR." Accessed: Oct. 10, 2025. [Online]. Available: https://www.up lift360.tech/tech/resin-removal-chemr

[160] C. Jenkins, and H. Doan, "Method of degrading a component of a fibre or fibre composite material", GB2637113A, 2023.

[161] T. Marasigan, "Uplift360 Awarded Innovate UK Smart Grant for Breakthrough Carbon Fibre Recycling Innovation." Accessed: Oct. 10, 2025. [Online]. Available: https://www.uplift360.tech/post/ uplift360-awarded-innovate-uk-smart-grant-for-breakthrough-carbon-fibre-recycling-innovation

[162] K. Shibata, and M. Nakagawa, CFRP Recycling Technology Using Depolymerization Under Ordinary Pressure, 2022.

[163] M. E. Kazemi, L. Shanmugam, D. Lu, X. Wang, B. Wang, and J. Yang, "Mechanical properties and failure modes of hybrid fibre reinforced polymer composites with a novel liquid thermoplastic resin, Elium®", *Compos. Part A Appl. Sci. Manuf.*, vol. 125, pp. 105523, Oct. 2019, doi: 10.1016/j. compositesa.2019.105523.

[164] M. Gebhardt, *et al.*, Reducing the raw material usage for room temperature infusible and polymerisable thermoplastic CFRPs through reuse of recycled waste matrix material, *Compos. B Eng.*, vol. 216, pp. 108877, Jul. 2021, doi: 10.1016/j.compositesb.2021.108877.

[165] J. A. Chiong, H. Tran, Y. Lin, Y. Zheng, and Z. Bao, "Integrating Emerging Polymer Chemistries for the Advancement of Recyclable, Biodegradable, and Biocompatible Electronics," *Adv. Sci.*, vol. 8, no. 14, Jul. 2021, doi: 10.1002/advs.202101233.

[166] P. Zamani, *et al.*, Assessing sustainability and green chemistry in synthesis of a Vanillin-based vitrimer at scale: Enabling sustainable manufacturing of recyclable carbon fibre composites, *Compos. Part A Appl. Sci. Manuf.*, vol. 179, pp. 108016, Apr. 2024, doi: 10.1016/j. compositesa.2024.108016.

[167] P. K. Dubey, S. K. Mahanth, A. Dixit, and S. Changmongkol, "Recyclable epoxy systems for rotor blades," *IOP Conf. Ser. Mater. Sci. Eng.*, vol. 942, no. 1, pp. 012014, Oct. 2020, doi: 10.1088/1757-899X/ 942/1/012014.

[168] F. Ferrari, C. Esposito Corcione, R. Striani, L. Saitta, G. Cicala, and A. Greco, "Fully Recyclable Bio-Based Epoxy Formulations Using Epoxidized Precursors from Waste Flour: Thermal and Mechanical Characterization," *Polymers (Basel)*, vol. 13, no. 16, pp. 2768, Aug. 2021, doi: 10.3390/polym13162768.

[169] Swancor, "Swancor Launched Recyclable Thermosetting Epoxy Resin." Accessed: Oct. 14, 2025. [Online]. Available: https://www.swancor.com/en/news/detail/Swancor-Launched-Recyclable-Thermosetting-Epoxy-Resin-EzCiclo-Leading-to-Zero-Carbon-Era

[170] G. Cicala, E. Pergolizzi, F. Piscopo, D. Carbone, and G. Recca, "Hybrid composites manufactured by resin infusion with a fully recyclable bioepoxy resin", *Compos. B Eng.*, vol. 132, pp. 69–76, Jan. 2018, doi: 10.1016/j.compositesb.2017.08.015.

[171] K. Jan, *et al.*, "Thermosets from renewable sources," in *Handbook of Thermoset Plastics*, Elsevier, 2022, pp. 679–718, doi: 10.1016/B978-0-12-821632-3.00011-7.

Kyle Pender, Patrick Sullivan
3 Thermal recycling of polymer composites

3.1 Introduction

Thermal recycling of fibre-reinforced polymer composites (FRP) involves volatilising the polymer matrix into low-molecular-weight polymers and gases such as carbon dioxide, hydrogen, methane, VOCs, or an oil/wax fraction [1–4]. These processes emphasise the reclamation of the fibre, fillers, and other inserts at the expense of the matrix in its raw material form. Key parameters that are considered during thermal recycling are the processing temperature and residence time which vary depending on the feedstock materials, feedstock size, atmosphere in the recycling process, accepted level of contamination, desired performance of recyclate, and desired rate of waste throughput. Depending upon the polymer matrix in the waste feedstock, the operating temperature of dedicated composite recycling processes typically lies within the range 400–700 °C [5], with temperatures up to 1,450 °C when used as feedstock in the cement industry [6]. Polymers such as unsaturated polyester resins (UPR) require lower temperatures, whereas epoxies or thermoplastics are operated at higher temperatures [5]. Thermal recycling can be broadly classified into two approaches: pyrolysis and combustion [7, 8]. The underlying concept of utilising heat to decompose the polymer fraction remains the same across each of these approaches; however, the methodology for heat generation and transfer, reactor type, and atmosphere deployed in the recycling process are factors which bring distinction to each of the classifications. Figure 3.1 shows the various approaches and classifications of composite thermal recycling technologies.

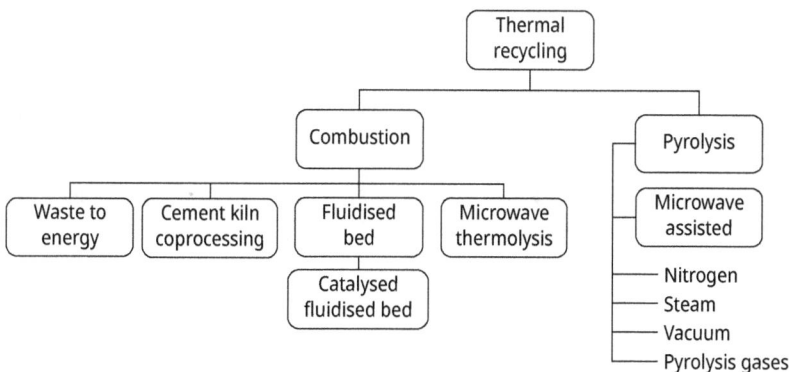

Figure 3.1: Polymer composite thermal recycling approaches and classifications.

https://doi.org/10.1515/9783110754438-003

Combustion recycling processes involve heating the waste composite in an atmosphere conducive for oxidation of the polymer matrix. Under these conditions, the polymer is ultimately valorised allowing the thermally stable, non-polymeric fractions in the waste to be recovered [9]. The presence of oxygen allows for rapid thermal decomposition of the polymer which can facilitate low residence times and high rate of waste throughputs and produce clean fibres with low surface contamination. Moreover, chemical energy in the polymer fraction is released during oxidation, which can directly heat the thermal process. Combustion processes have the drawback of utilising the entire polymer as fuel, opposed to recycling in material form. Examples of combustion recycling processes for composite waste include the fluidised bed [10], the catalysed fluidised bed [11], and cement kiln co-processing [12].

Pyrolysis recycling uses indirect heat, in an oxygen-starved environment, to thermally decompose the polymer fraction in waste composites. Limiting oxidation allows for the potential recovery of not only reinforcement fibres (as well as fillers and inserts when present) but also the polymer fraction in the form of hydrocarbon gases, liquids (oils), and solids (wax, char) [1]. Contamination on the surface of pyrolysed fibres is commonly observed in the form of carbonised matrix residues. This has led to secondary oxidation steps often being implemented to facilitate the removal of the thermally stable char and yield contaminant free fibres. Pyrolysis is a rather more complex process than combustion and is less tolerant to mixed or unknown polymers waste streams. This is compounded when attempting to recover material value from the recover hydrocarbon fractions (beyond simply using as fuel), where consistency in waste feedstock is conducive to producing a commercially attractive secondary raw material.

Lastly, microwave-assisted thermal recycling involves targeting microwave irradiation onto the waste composite [2]. This produces the heat necessary to thermally decompose the polymer fraction and liberate the fibrous and other non-organic fractions from the waste. This process is not necessarily distinguished from pyrolysis and combustion approaches; rather it is characterised by the approach taken to heating the waste. This process is generally utilised in an oxygen-starved environment (named microwave-assisted pyrolysis); however, it has also been studied as a method for heating in combustion-based recycling. The proposed advantage of this method for composite recycling is that the reclaimed fibres retain higher strength using a low energy and fast heating process; however, unlike other composite thermal recycling technologies, this method has only been demonstrated at the laboratory scale.

3.2 Thermal recycling processes

While composite thermal recycling methods all attempt to reclaim the fibrous reinforcement fraction, the method used to remove the polymeric matrix fraction differs widely and has disparate products and potential uses/routes to market. There are several classifications of thermal processes used for plastic recycling that are relevant for understanding the current and future strategies for polymer composites recycling. Figure 3.2 provides a schematic of these different approaches showing the classification of products expected from each. Combustion (or incinerating) is the most basic example of thermal decomposition process producing fully oxidised compounds (such as CO_2 and H_2O), which is done at large scale in waste to energy facilities [13]. For neat polymers this is not strictly considered a recycling technology; however, it can be used to recycle the fibre fraction in composites.

Gasification involves heating the polymer to a very high temperature (>1,000 °C) in low levels of oxygen which results in partial oxidation of the polymer into syngas (H_2, CO and CO_2, and CH_4 and N_2 mixture), which can then be used in the production of a range of chemicals for plastics production, fuels, and fertiliser [13, 14]. Gasification can co-process mixed polymer waste streams which is advantageous for multi-material/contaminated composite waste streams (such as wind turbine blades); however, the temperatures used are likely prohibitively high to extract value from reclaimed fibre fractions.

Pyrolysis is a rapidly growing technology, demonstrating capability in converting waste polymers to valuable chemical feedstocks and fuels [15]. The basic process involves thermally degrading material(s) in a starved oxygen environment. When applied to polymeric materials, this results in thermally cracking the polymer into simple hydrocarbons, oligomers, and monomers. The hydrocarbon vapours can be condensed and collected as waxes, oils, and gas, and the distribution of which can be controlled by adjusting process time and temperature [16]. Pyrolysis products can be processed using conventional refining technologies to produce the feedstock for polymers production or used directly as a fuel. Typical plastic feedstocks for pyrolysis include polyolefins, polystyrene, and PMMA [17–19]. Pyrolysis can process contaminated and mixed-polymer waste streams which, alongside the lower operating temperature, make pyrolysis an attractive and heavily research approach to polymer composite thermal recycling. The mitigation of direct emissions and the opportunity to reclaim the polymer in usable material form is a major advantage of pyrolysis over incineration-based processes (such as is used in traditional waste to energy facilities).

Depolymerisation (or chemolysis) is the reverse of polymerisation and typically yields monomer or oligomers close to the quality of virgin counterparts, meaning these products are of higher value and can be used directly in the production of new polymers. Depolymerisation can only be used with condensation polymers (such as PET and polyamides) and typically cannot be used for the decomposition of addition polymers (such as polypropylene (PP), PE, and PVC) or thermosetting polymers widely used in composites [20].

As discussed above, technology developments in thermal recycled of composite materials have focussed on combustion and pyrolysis-based processes.

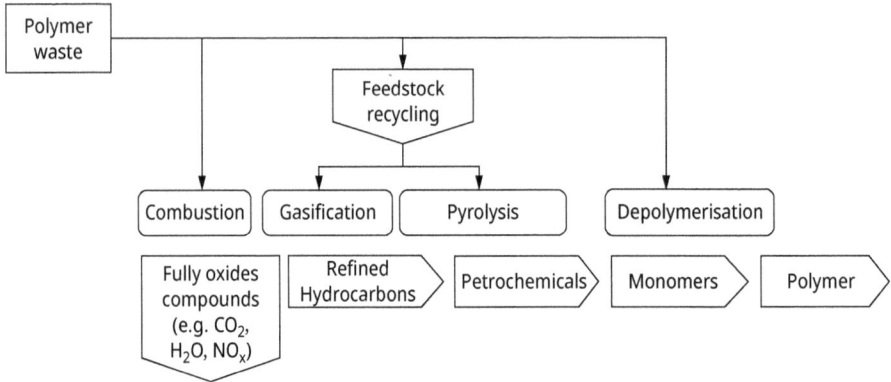

Figure 3.2: Thermal-based processes used to decompose polymers with relevance to composite recycling.

3.2.1 Recycling using pyrolysis

Amongst different methods of recycling of composites, pyrolysis has progressed to industrial level in recent years [21]. Pyrolysis recycling of FRP involves heating waste to 300–700 °C in low or no oxygen environment [22], which can be done in either a batch or continuous process, depending on the type of pyrolysis method employed. The pyrolysis process can theoretically allow for the recovery of the polymer matrix as well as the reinforcement fibres. The liquid and solid material can be recovered and used as feedstock for further chemical processes and the gas can be utilised in combustion processes and/or to heat the process.

The low oxygen environment limits the oxidation of carbon fibre (CF) reinforcements, which is key to extracting value from carbon fibre-reinforced polymer (CFRP) recycling. The resident time and process temperature are tailored to both the wastes polymeric composition as well as the thickness of the laminate being recycled. Insufficient processing parameters can result in a layer of pyrolytic carbon formation on the fibres [1, 23, 24], which can be removed by additional processing in oxidative conditions [25–27]. This secondary combustion step can cause negative effects on mechanical characteristics in the case of recycled carbon fibre (rCF), resulting in reduced tensile strength and elasticity [27–29]. This can be minimised, however, through optimisation of the operating conditions such as pyrolysis and oxidation temperatures, residence time, and reaction atmosphere. This can produce fibres in a suitable condition to be used as feedstocks in a secondary material with sufficient fibre/matrix adhesion [25–27]. Beyond the physical size of the reactor or furnace used, there is no limitation on the feedstock size for pyrolysis; therefore "long" fibres can be recovered. A diagram of a typical pyrolysis process for composites recycling is given in Figure 3.3.

For pyrolysis, typical atmospheric conditions include nitrogen, vacuum, carbon dioxide, helium, argon, recirculated pyrolysis product gases, and superheated steam.

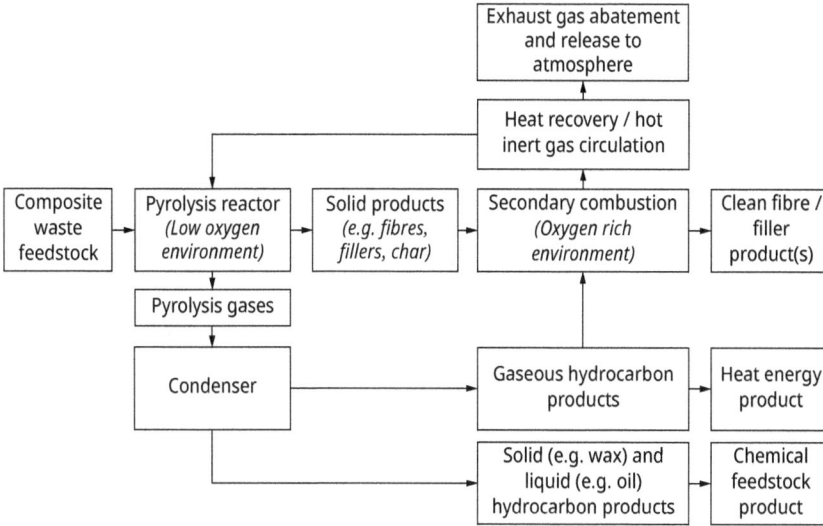

Figure 3.3: Diagram of a typical pyrolysis process.

Nitrogen purging is a common technique used to create an inert atmosphere in batch pyrolysis and lab-scale academic studies. Early studies by Onwudili et al. [30] investigated pyrolysis of CFRP (polybenzoxazine resin system) under nitrogen atmosphere in a batch stainless steel autoclave between temperatures 350 and 500 °C and reaction times of up to 60 min. Solid products were found to be mostly rCF with high yield of up to 98 wt.%; liquid products consisted of 15–20 wt.% water combined with organic oils and gas products contained mainly carbon dioxide and some hydrocarbon gases. The proportion of the combustible gases had calorific values of up to 35 MJ/m^3 which is approximate to natural gas in energy density [30]. In 2016, Onwudili et al. [28] used a semi-batch reactor with nitrogen atmosphere at 500 °C to investigate the mechanical characteristics of both recycled CFRP and glass fibre-reinforced polymer (GFRP) waste. Based on SEM micrographs it was determined that, with pyrolysis alone, the surfaces of both fibre types were covered with the residues from the degraded polymers, which could be removed by a subsequent oxidising process.

Steam pyrolysis is another pyrolysis approach that has been applied to FRP recycling, which refers to the thermal decomposition of organic compounds that are heated to high temperatures in the presence of steam. In steam pyrolysis recycling of composites, a superheated steam atmosphere has been used in an attempt to improve heat transmission, accelerate the thermal breakdown of the polymer matrix, and facilitate the removal of oxygen from the pyrolysis reactor [31]. Shi et al. [32] reported significant levels of polymer residue on both rCF and recycled glass fibre (rGF) recycled following pyrolysis of UPR composites at 340–370 °C for 30 min. Kim et al. pyrolysed CF-reinforced epoxy composites with superheated steam at 550 °C in a fixed bed reactor, finding that after 30 min, rCF were heavily covered in polymer residue. The introduction of hot air was shown to yield cleaner rCF and using FT-IR and XPS it was concluded that oxygen-containing functional groups increased after pyrolysis in steam and air which improves the chemical activity between the rCF and the resin matrix [33]. Kim et al. investigated a two-step pyrolysis technique, consisting of low-temperature decomposition in carbon dioxide at 400 °C, followed by high-temperature decomposition using superheated steam method at 700 °C. Using SEM, contaminant-free rCF were obtained after more than 40 min of superheated stream treatment [34]. Other pyrolysis atmospheres that have been shown to produce clean recycled fibres from FRP include combined nitrogen and superheated steam (temperature not specified) [35] as well as combined helium and superheated steam (pyrolysis temperature of 600–700 °C) [36]. While these authors claim to be using pyrolysis, the exact mechanism for polymer decomposition, and whether hydrolysis is indeed occurring remains unclear.

An atmosphere of reduced pressure can produce cleaner fibre from polymer composite waste. Greater diffusion of product molecules from the decomposing solid phase is possible under vacuum meaning they are more rapidly eliminated from the reaction zone [37]. A novel two-stage process of vacuum pyrolysis and centrifugal separation was found to successfully enable combined recovery of solder and organic materials from composite waste printed circuit boards (WPCBs) [38]. The results of centrifugal separation indicated that the separation of solder from WPCBs was complete when WPCBs were heated at 240 °C. The results of vacuum pyrolysis showed that the yield of pyrolysis products was dependent on the type of WPCBs, with values ranging 69.5–75.7 wt.% residue, 20.0–27.8 wt.% oil, and 2.7–4.3 wt.% gas. The pyrolysis residues contain various metals, glass fibres (GF), and other inorganic materials, which would require further processing prior to reuse [38].

The distribution of the pyrolysis products is dependent on the polymer type and treatment temperature [8]. The gas product accounts for up to 20 wt.%, with the majority being CO_2 and CO, giving a relatively low calorific value of around 15–20 MJ/kg [3, 22, 39]. The liquid fraction accounts for 5–50 wt.% of the recovered material with a high calorific content of 30–40 MJ/kg, which is in the range of fuel oil [8, 22, 39]. The solid product typically accounts for 30–95 wt.%, which comprises the rGF, mineral filler, and char residue [3, 8, 22, 39]. Table 3.1 compiles processing parameters and material output data from studies recycling GFRP and/or CFRP using pyrolysis.

Polyolefins have been shown to make an ideal feedstock for pyrolysis as they can degrade into valuable hydrocarbon products [40]; therefore pyrolysis may be ideally suited to thermally recycle filled/reinforced thermoplastic polyolefins (e.g. polypropylene and polyethylene) or elastomers (e.g. ethylene propylene diene monomer). As well as polyolefins, closed loop pyrolysis recycling of CF-reinforced polymethyl methacrylate (PMMA) has been demonstrated. Bel et al. [41] investigated pyrolysis of CF-reinforced PMMA (Elium, Arkema), demonstrating that MMA monomer could be collected at high enough purity to re-process directly back into PMMA and produce a second life CFRP with comparable mechanical performance to virgin counterpart.

Uncured CFRP prepreg (polybenzoxazine) w/ LDPE film	Muffle furnace	2 kg	58.4–61.5	28.5–30.7	10–10.9	350–700 °C	Air, 500 °C, 30–180 min	López et al. [27]
GFRP (UPR)	Muffle furnace	4 kg	68	24	8	550 °C, 180 min	Vitrification, 1,450 °C, 120 min	López et al. [44]
GFRP WPCBs (epoxy)	Pilot scale fixed bed (500 mm diameter)	5 kg	78.3	14.5	7.2	Vacuum, heat up 10 °C/min, 500 °C, 1 h	–	Li et al. [42]

While pyrolysis technology has been developed to commercial scale to process neat plastic waste, there are several technical challenges that remain. Plastics collected from disparate waste streams are heterogeneous in nature and yield a mixture of different hydrocarbon products of various chain lengths. Even relatively small dosing of other polymer types (such as PVC) in the feedstock can produce undesirable chemical products which can be detrimental to the process itself as well as contaminate and deteriorate the quality of oil products. This is of particular concern for feedstocks containing halogen flame retardants, such as brominated flame retardants (e.g. tetrabromobisphenol A) which are widely used in composites industries; acidic by-products can cause corrosion of the pyrolysis reactor as well as render the oil halogenated [42, 43]. Moreover, applications for the use of pyrolysis products of widely used thermoset resins used in FRP, such as epoxy resins, UPR, and vinyl ester resins have yet to be reported.

Upon analysing the products from pyrolysing GF-UPR composite (which comprised benzene, toluene, ethylbenzene, styrene), Giorgini et al. [1] concluded that low content of sulphur and halogens (most likely due to absence of halogen flame retardants), together with the absence of heavy metals, make the obtained pyrolysis oils suitable for use as fuel without requiring any further purification process. Cunliffe and Williams used gas chromatography-mass spectrometry to analyse the composition of pyrolysis products collected from GF-UPR waste. Styrene and phthalic anhydride accounted for 19 and 27 wt.% of the total condensable product yield, respectively, which each have multiple uses in the polymer sector [4]. Styrene is used in the production of polystyrene, styrene–butadiene rubber, acrylonitrile butadiene styrene (ABS) and UPR, whereas phthalic anhydride can be used as a cross-linking agent for epoxy resins and as a catalyst for the polymerisation of olefins. Cunliffe and Williams also observed oxygenated and bromated species in the products (2-bromopropiophenone), which were attributed to the UPR component of the copolymer resin and flame retardant in the waste, respectively [4]. Due to the high oxygen content, the gross calorific value of the collected oil was low relative to conventional hydrocarbon liquid fuels but was sufficiently high to make it a viable fuel source.

Aside from aromatics, pyrolysis oil from other widely used polymer composite matrices can contain significant quantities of oxygenated species [3, 33, 42, 44, 45]. Ren et al. and Hiltz [2, 46] analysed the liquid products from cured epoxy and vinyl ester polymers, respectively. Epoxy was found to produce several aromatics, including benzene and a range of phenolic compounds (phenol, p-isopropyl phenol, bisphenol A, methyl tetrahydrophthalic anhydride) [2, 42, 45]. Similarly, pyrolysis of bisphenol A–based vinyl ester resin was found to produce several degradation products that are characteristic of the bisphenol A portion of the resin including phenol and methyl-substituted phenols [46]. Table 3.2 gives the temperature range and pyrolysis product species for thermoplastic and thermoset polymers commonly used in FRP.

To re-introduce the pyrolysis liquids as a material in the petrochemical industry (opposed to simply as a fuel), these mixed hydrocarbon products could be co-fed into

Table 3.2: Typical pyrolysis characteristics and product species of common polymer matrices (reproduced from [37]).

Polymer	Pyrolysis temperature (°C)	Char (%)	Character of liquid
Thermoplastic polyurethane	300–370	5	Aliphatic ester/aromatic
Polyester	370–460	26	Aromatic
Epoxy	370–460	15	Alkylphenol
Polybutylene terephthalate	370–430	3	Aromatic acid/aliphatic
Polyethylene terephthalate	400–460	11	Aromatic ester
Nylon 6	430–490	0	Aliphatic
Nylon 12	440–490	0	Aliphatic
Polypropylene	450–500	0	Aliphatic
Polycarbonate	480–570	22	Alkylphenol
Phenolics	450–580	30–50	Methylphenol
Polyphenylene oxide	520–580	54	Phenol
Polyether ether ketone	560–620	55	Phenol

existing crude oil refinery units. Where phenolic compounds are present, the high oxygen content (typical oxygen content of crude oil is below 1 wt.%) and acidity could cause refining issues with existing infrastructure [47]. This includes the higher boiling point of oxygenate compounds compared to hydrocarbon with the same carbon number as well as the presence of various reactive oxygen-related functionalities allowing thermal polymerisation and high coke rate [47]. Pre-processing of pyrolysis products is therefore likely needed, especially when processing multi-source FRP waste streams, which is yet to be demonstrated technically or economically feasible at scale.

Unlike combustion, pyrolysis is an endothermic process, requiring energy input to crack the polymer matrix [15]. Due to the high primary energy demand of virgin CF production [48], the energy demand during pyrolysis recycling of CFRP is significantly lower than virgin CF production and is therefore a strong environmental incentive to use this technology (regardless of if the matrix fraction is reused as fuel). On the contrary, energy required for virgin GF production is significantly lower and therefore requires a more efficient process to achieve a favourable energy balance [49].

3.2.2 Recycling using combustion

There are currently three combustion-based solutions for composites at end-of-life, with technologies across the TRL scale from lab-scale feasibility demonstration, through to commercially available and actively exploited solutions. These classifications include (1) incineration (either with or without energy recovery), (2) cement kiln co-processing, and (3) the combustion-fluidised bed. While energy can be recuperated from polymer fractions in composite waste, incineration with energy recovery is not considered recycling, given that there are no restorative material flows through the process. Therefore, this route is omitted from this section, which will focus on cement kiln co-processing and the fluidised bed, as these technologies have been demonstrated to be capable of extracting material value from composites waste streams.

3.2.2.1 Cement kiln

Cement co-processing is commercially available for processing large volumes of GFRP waste, particularly scrap wind turbine blades. Environmental gains can be made by using waste GFRP to substitute both fuel and raw materials (CaO, SiO_2, $Al2O_3$, Fe_2O_2) in producing clinker. In this process the mineral components are reused in the cement; however, the structure and function of the GF is lost during the process, which from a waste hierarchy perspective may be less preferred [28, 57, 58]. The polymer fraction in the GFRP is recovered as fuel (containing about 12 MJ/kg of GFRP waste) to heat the energy-intensive calcification process.

Co-processing in a cement kiln requires that GFRP material to be shredded (<40 mm-sized pieces) and then mixed with solid recovered fuel (SRF), an alternative fuel made from mixed dry waste that are challenging to separate and would otherwise go to landfill/incineration [6]. SRF is widely used as a substitute for fossil fuels in the cement industry, which must be prepared in compliance with specification EN 15,359 [59]. The polymer in the GFRP fuels the process and helps to bring the cement kiln temperature above 850 °C, replacing a portion of fossil fuels that would otherwise be used in the process (such as coal or petroleum coke). The temperature in the kiln must then be brought to beyond 1,450 °C using fossil fuels, to enable the alumina-borosilicate in the GF fraction and the calcium carbonate to both calcify, turning into alumina, silica, and calcium oxide, all of which are key components of Portland cement [6]. Through this process, all the composite waste is used (either as a material feedstock or fuel) and nothing goes to landfill.

The proportions of mineral components in the GF are not perfectly matched with the raw material requirements for clinker production (approximately 20% SiO_2, 10% Al_2O_3 + Fe_2O_2, and 70% CaO [60]); therefore GFRP waste must be added in certain proportions in order to meet the clinker requirements. The requirements for clinker feedstock materials also vary geographically and are highly dependent on the composition of local resources which the plant has been designed to absorb. As such, analysis of the waste composition must be conducted prior to establishing GFRP as a new feedstock in clinker production and added in proportions that meet the regional needs of the kiln. In addition to this, it has been reported that only 10% of the fuel input into a cement kiln should be substituted with polymer composite material, due to the presence of boron in the composite waste which can slow cure time [8]. When introduced correctly, however, it has been established that the inclusion of the solids does not affect the ultimate strength of the cement [8, 61].

As with mechanical recycling, this method may be seen as a down-cycling solution as it does not utilise the reinforcement potential of GF. Despite this, an environmental impact analysis conducted by Quantis USA, a sustainability consulting group, found that compared to traditional cement manufacturing, blade recycling enables a 27% net reduction in CO_2 emissions from cement production and 13% net reduced water consumption. In addition, a 7 ton wind blade recycled through the process enables the cement kiln to avoid consuming nearly five tons of coal, 2.7 ton of silica, 1.9 ton of limestone, and nearly 1 ton of additional mineral-based raw materials [62].

3.2.2.2 Combustion-fluidised bed

A fluidised bed is a phenomenon that can occur when a quantity of a solid particulate substance is placed under suitable conditions to cause a solid-fluid mixture to behave as a fluid. This is typically achieved by pumping pressurised fluid into a bed of particulates. The resulting medium then has many properties and characteristics of normal fluids, such as the ability to freely flow under gravity, or to be pumped using fluid technologies. Fluidised beds have several cross-sector applications, such as fluidised bed reactors (chemical reactors), fluid catalytic cracking, fluidised bed combustion, solids separation, heat and/or mass transfer, and more [63].

The University of Nottingham pioneered this process for recycling GFRP waste in the early 2000s, and since then the combustion-fluidised bed process has been demonstrated to successfully thermally recycle CFRP [10, 64, 65]; while overcoming industrial challenges such as scalability, operation continuity, and contaminant sensitivity, recovering clean fibre and processing dissimilar polymers [8, 24]. For FRP recycling, fluidised bed reactor contains silica sand that is heated by a stream of pre-heated air (schematic shown in Figure 3.4). The composite is automatically fed into the reactor and polymeric fraction is rapidly combusted in the fluidised sand, due to the presence of oxygen in the reactor, high heat transfer rates, and mechanical abrasion. After liberation from the matrix, fibres and filler materials are carried by the stream of hot gases out of the reactor and separated using solid-gas separation techniques. The gas stream is subsequently introduced into a secondary combustion chamber where it is fully combusted at up to 1,000 °C to produce clean flue gas [64]. Heat energy can be recovered from the combustion chamber to preheat incoming fluidisation air entering the reactor. The typical temperature within the fluidised bed reactor is 450–550 °C, depending on the polymer within the waste [9, 11, 64–66].

The novelty in using the fluidised bed for composite waste is the ability to process continuously while simultaneously separating and collecting non-combusted constituents. This is particularly important for filler-laden composites such as bulk-moulding compound (BMC) and sheet moulding compound (SMC), which are a significant contribution to annual FRP waste volumes and contain up to 50 wt.% mineral fillers [67]. Pickering et al. [10] demonstrated the functionality of a multi-stage separation

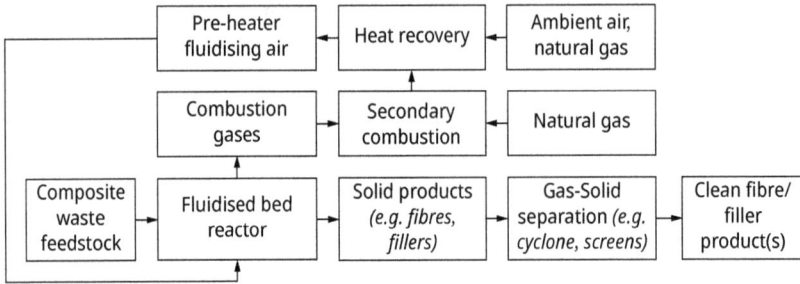

Figure 3.4: Schematic of fluidised bed thermal recycling process.

approach to reclaim both rGF and calcium carbonate filler independently and continuously from waste SMC, filament wound GFRP pipe, and GFRP sandwich panels using the fluidised bed. Metallic inserts often present in real-world composite scrap sink to the bottom of the sand bed and can in theory be recovered without disrupting the process by utilising bed particulate recirculating methods [10]. This has yet to be demonstrated for composite recycling; however, it is a process used in commercial fluidised bed reactors to remove contamination accumulation over time.

Unlike other batch or belt-driven processes, the fibre residence time in the fluidised bed reactor is not an independent operating parameter of the system. Rather, it is self-regulatory, and fibres are liberated from the outer surfaces of the waste material as the exposed layer of polymer is thermally decomposed [64]. After release from the matrix, the fibre is free to leave the sand bed (called entrainment), driven by the continuously circulating gas flow through the system. Surface contaminants on the fibre make the fibre heavier, increasing its terminal velocity, which requires a greater gas velocity to remove the fibre from the reactor freeboard. This phenomenon means that gas flow rate through the reactor can be optimised to limit the contaminated fibre entrainment, allowing only the lighter weight, residue free fibres to be collected. Residence time in the reactor is therefore minimised while ensuring recycled fibres are clean.

While this is a combustion process, it has yet to be demonstrated to be self-sustaining and currently requires additional heat input in the form of natural gas introduced in the off-gas combustion chamber. While heat energy is continuously introduced into the reactor in the form of oxidised polymers in the FRP feedstock, a relatively high air mass flow rate entering the reactor is required to fluidise the sand [68]. The composite feed-rate is limited to the residence time of fibres already in the reactor, with over packing resulting in disturbances in the fluidisation dynamics of the sand as well as fibre-fibre agglomeration [69]. As a result, the reactor runs oxygen rich and relatively large volumes of fluidisation air need to be continuously preheated to maintain the desired reactor temperature. High efficiency gas-gas heat exchangers have been used to minimise the use of natural gas in the system [70].

The fibre length entering the fluidised bed reactor must also be controlled and plays a key role in determining the reactor loading rate (the rate at which FRP can be fed into the fluidised bed reactor). Prior to feeding into the fluidised bed, the waste composites must undergo size reduction to 10–25 mm using a shredder, granulator, or hammer mill [64, 66]. A systematic analysis of reactor loading rates for FRP recycling in the fluidised bed has not been carried out in the literature; however, it is understood that it is limited by fibre entanglement and formation of fibre agglomerates within the fluidised bed [69]. The mechanism for how these agglomerates form is currently not fully understood; however, fibre length and residence time in the fluidised bed play a role in their prevalence. As such, the degree of downsizing of FRP waste prior to recycling must be adequate for the desired reactor loading rate to avoid fibre agglomeration. Other operating conditions such as reactor temperature and fluidisation velocity will also influence fibre residence time within the bed so should be tuned accordingly [71].

A fluidised bed process for recycling FRP has not been commercialised; however, a 50 ton/year pilot scale system has been developed by the University of Nottingham, used for recycling CFRP [69, 70]. A 350 ton/year pilot scale fluidised bed for GFRP recycling is underdevelopment, led by researchers at the University of Strathclyde [72]. Pickering et al. [10] conducted a study demonstrating a fluidised bed recycling process could potentially financially break-even at a capacity of around 9 kton of scrap GFRP per annum. A recent economic analysis by Meng et al. concluded that recovery of rCF using the fluidised bed can be achieved at less than $5/kg at a relatively low plant capacity of 1 kton of CFRP waste per annum. This presents a significant cost reduction compared to virgin CF counterparts [73].

Variations on the pyrolysis fluidised bed are also widely researched as a method for cracking mixed, neat polymer waste, in what is typically termed "chemical recycling" of plastics. Fluidised bed pyrolysis processes exhibit process continuity, short residence time, high contact surface, and superior heat and mass transfer (5–10 times higher heat transfer compared to indirect or direct heated melting vessels, shaft reactors, and rotary kilns [74, 75]). The reactor temperature and fluidising gas used (e.g. nitrogen, steam, and recycled pyrolysis gas) are key parameters used to optimise the process and can produce different products from plastic waste. Enabling low residence time, and fast pyrolysis of the polymer, can result in less secondary reactions and produce fewer side products [76]. Depending on the polymer and processing parameters, fluidised bed pyrolysis can produce mainly waxy products, oil, gas, or monomers [74, 77]. Fluidised bed pyrolysis recycling of composite waste has yet to be demonstrated; however, it may be an area of interest for future development of the combustion fluidised bed. This approach may retain the advantages that combustion fluidised beds have over the batch/static pyrolysis processes utilised for composite recycling to date, while enabling recovery of the polymer in material form.

3.2.2.3 Catalysed combustion-fluidised bed

The catalysed combustion-fluidised bed recycling process is a variation on the standard combustion-fluidised bed process which aims to (1) reduce the temperature required to recycle FRP and/or (2) facilitate an increase in composite waste throughput [66]. This is achieved by introducing a catalyst into the system to reduce the thermal stability and accelerate the combustion of the polymeric matrix [78]. This has the potential of reducing both the energy consumption of the recycling process and thermal damage sustained by the fibres during recycling [11]. To date, this technology has only been applied to GFRP recycling. Pender et al. [78] investigated a range of metal oxide catalysts in the combustion-fluidised bed-recycling process, reporting on the impact on rGF reclamation efficiency, recycling temperature, rGF cleanliness, and rGF tensile strength. Metal oxides encompass a widely used category of solid catalyst with transition metal oxides being utilised in many organic reactions [79]. The following typical redox mechanism describes the catalytic oxidation reactions on metal oxides:

$$Me - O + Red \rightarrow Red - O$$

$$Me + Ox - O \rightarrow Me - O + Ox$$

Initially a reductant (Red) reduces the metal oxide surface (Me-O). It returns to its original state after re-oxidation by an oxidant (Ox-O) [79]. The result of this two-stage reaction is the transfer of oxygen from one substance to another [80]. Oxidation of organic substances occurs as a result of the metal oxide donating a lattice oxygen, producing a vacancy on its surface [81]. The air stream in the fluidised bed process can therefore be responsible for re-oxidising the surface oxygen vacancy, facilitating a continuous redox cycle. Figure 3.5 gives the obtained rGF yields when recycling GF-epoxy in the fluidised bed at 400, 450, and 500 °C and compares the results with and without integrated copper oxide (CuO) catalyst [11]. The application of CuO significantly improves the yield efficiency of rGF, particularly at lower temperatures. When recycling at 400 °C the yield increases from 6% to 40% when using CuO. As would be expected, the yields converge at higher temperatures since the epoxy rapidly thermally decomposes without the need for an oxidising catalyst. Pender et al. [11] concluded that CuO can aid epoxy decomposition within the fluidised bed, increasing fibre liberation, in turn, improving yield efficiency at lower operating temperatures.

3.2.3 Microwave-assisted thermal recycling

Microwave heating is a growing technology used in material processing due to its advantages of rapid, uniform, and selective heating [82, 83]. Microwave heating involves heating through absorption of electromagnetic energy within a material, the heat is generated from the inside and conducts to the entire volume [84, 85].

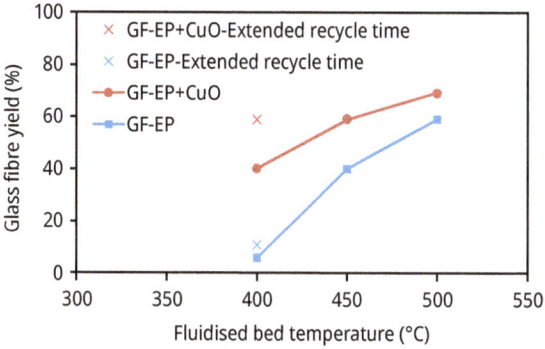

Figure 3.5: rGF yield efficiency of GFRP recycled in the fluidised bed at various temperatures showing significantly higher yield when using oxidation catalyst (produced using data reported in [11, 66]).

3.2.3.1 Microwave-assisted pyrolysis

Microwave-assisted pyrolysis uses microwave radiation to replace traditional pyrolysis heating sources, with the aim to increase efficiency and reduce energy demand of the process. As with traditional pyrolysis, microwave-assisted pyrolysis of FRP is conducted in an oxygen-starved environment to limit oxidation of the polymer [5, 86, 87]. An early feasibility study by Lester et al. [87] used a 3 kW multimode microwave applicator to recycle epoxy-based CFRP, concluding that the polymer could be separated from the rCF using the technology and that the technique may offer a useful route to recovery of long rCF, together with a pyrolysed polymer vapour. The rCF were reported to be relatively clean, and strength retention was superior to those recovered using fluidised bed volatilisation.

Obunai et al. [88] recycled rCF GF-epoxy waste by irradiating microwaves under different atmospheres. The effect of the atmosphere (argon, nitrogen, and air) and field intensity of irradiated microwaves on the efficiency of extraction of rCF was investigated. It was concluded that irradiation power of 700 W for 5 min was sufficient to recover clean rCF. Under SEM, pitting on the surface of fibres recycled in the air environment was observed (whereas smooth surfaces were observed when performed under nitrogen or argon), likely a result of surface oxidation. The temperature of the specimen during recycling was not reported, making comparison to other studies challenging. Obunai et al. [88] proposed a three-stage mechanism for rCF extraction using microwave irradiation. First, the CF in CFRP is joule heated by induced current by microwave irradiation. Secondly, gasification of resin is promoted by the heating of the embedded CF. And, finally, the gasified resin is further decomposed by spark glow plasma induced by spark glow discharge between CF [88].

Akesson et al. [89] recycled rGF from GFRP wind blade waste using the microwave pyrolysis method. This was carried out using a total power of 3 kW, at temperatures of 300–600 °C in a nitrogen environment. TGA showed that the pyrolysed fibres had an organic residue content of about 3–8 wt.%. Under SEM the fibres appeared to be coated with some residual material, assumed to be of carbonaceous nature, which may be derived from UPR that had not completely degraded during the pyrolysis. Lower mechanical properties of composites made with the rGF was observed and explained by the surface contamination restricting fibre-polymer adhesion and suggesting that secondary heating in oxygen could be implemented to remove the char, similar to what is utilised in convention pyrolysis processes.

Jiang et al. [90] investigated the use of microwave irradiation in nitrogen to recycle CF-epoxy. A temperature range of 400–600 °C and residence time of 30 min was selected for the trials. SEM micrographs of the rCF after pyrolysing at 400 °C showed a large amount of pyrolytic residuals visible on the fibre surface. At 500 °C, the rCF surface was much cleaner, but some surface residue remained. At the highest temperature of 600 °C, very low levels of residue were present; however, signs of damage appeared as cavities on the rCF surface, likely because of oxidation despite the low oxygen environment. Regardless, it was concluded that microwave irradiation was a flexible, easily controlled, efficient technique to reclaim the rCF from CFRP composites and that the rCF can be directly used as reinforcement in a new polymer (PP and Nylon) using traditional composite processing technology [90].

3.2.3.2 Microwave thermolysis

A lesser investigated thermal recycling approach for FRP is "microwave thermolysis", which utilises microwave heating of the FRP, but does so in an air atmosphere to enable fibre recovery through oxidation of the polymer matrix. Deng et al. [82] investigated microwave thermolysis for CF-epoxy recycling, using a high-temperature microwave furnace and compared the recycling parameters and rCF against those heated in a traditional muffle furnace. It was found that the epoxy matrix completely decomposed, and rCF could be collected using microwave heating when the temperature reached 450 °C. Observation of rCF oxidation was made using SEM when heated to 500 °C and was concluded that 450 °C was optimal for fibre reclamation. Compared to traditional thermo-oxidative decomposition in the furnace, it was concluded that microwave thermolysis is a faster and more efficient method that requires less energy, reduced residence time by 57%, and can increase fibre recovery rate by 15% [82]. The surface of the rCF recovered using microwave heating were cleaner, smoother, and contained less epoxy resin compared to conventional heating.

3.3 Mechanical performance of thermally recycled fibres

The terms "thermal conditioning" and "thermal recycling" will be used here; for clarity, these terms will now be defined. Thermal conditioning involves directly exposing fibres to an environment with an elevated temperature. In this case, the fibres are not within a matrix material when subjected to an elevated temperature and are done so mimic the conditions of a thermal recycling. Thermal recycling involves exposing FRP to an environment with an elevated temperature to thermally decompose the polymer matrix and liberate the reinforcement fibres.

3.3.1 Thermally recycled glass fibre

Extensive research has been carried out to investigate the effect of temperatures approximating those used in composite thermal recycling on GF mechanical performance. It is concluded by numerous authors that thermal conditioning [66, 78, 91–97] and thermal recycling [9, 11, 39, 64, 71, 89, 98, 99], processes can cause a significant reduction in GF strength. It can therefore be stated that exposure to elevated temperatures leads to "thermal weakening" of GF. Both temperature and time of exposure influence the extent of fibre weakening. GF strength loss was found to increase with temperature when thermally conditioned and thermally recycled. It was observed that strength loss also increases with exposure time, until a constant minimum value is reached [93, 100]. An increase in conditioning temperature is associated with more rapid GF weakening since this asymptotic minimum is attained faster at higher temperatures. Figure 3.6 shows the effect of both temperature and time on GF residual strength.

The mechanism of thermal weakening of GF is still under discussion in the literature, and there is no consensus on the physical change(s) that occur in GF that can account for the strength loss [101]. The proposed mechanisms can typically be split into two categories, bulk or surface phenomenon, and are discussed in detail by Thomason et al. [101] in relation to composite thermal recycling. Feih et al. [102] proposed that the strength loss is due to an increase in flaw severity on the surface of thermally conditioned GF. Indeed, the strength of GF is typically considered to be controlled by surface flaws [101, 103]. Feih et al. [102] introduced an "artificial" surface flaw on the surface of a silane-sized GF using an ion beam. It was found that the strength of GF with surface flaws induced closely approximated those of thermally conditioned GF. Furthermore, fracture surface analysis and fracture toughness modelling showed that the properties of the bulk structure were unchanged following exposure to elevated temperatures. It was concluded that the surface flaws only needed to grow between 180 and 400 nm (depending on flaw geometry) during thermal conditioning.

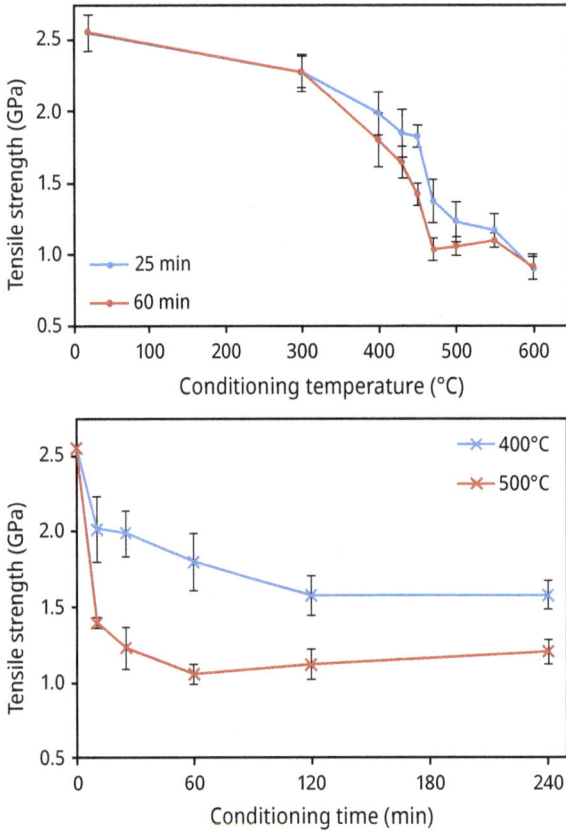

Figure 3.6: Effect of both temperature and time on GF residual strength [66]. Top – GF thermally conditioned for 25 and 60 min at various temperatures from 300 °C to 600 °C. Bottom – GF thermally conditioned at 400 °C and 500 °C for various durations up to 240 min.

Feih et al. [102] suggest that flaw growth occurs during thermal recycling by water molecules (from sizing or atmosphere) diffusing into the GF structure and reacting with stressed siloxane bonds at the crack tip. The surface flaw theory is bolstered by works carried out by Sakka [104], Yang et al. [91], and Pender and Yang [9] who demonstrate that thermally weakened fibres can be partially re-strengthened by surface etching.

The surface of rGF recycled in the fluidised bed was further studied under high magnification SEM, as seen in Figure 3.7. The presence of apparent damage was ubiquitous on the surface of rGF in the form of scratches and depressions. It cannot be concluded whether the damage is induced thermally or mechanically during recycling. No such features were observed on the surface of GF simply thermally conditioned within a furnace at the same temperature, suggesting that the damage is a result of mechanical attrition during recycling.

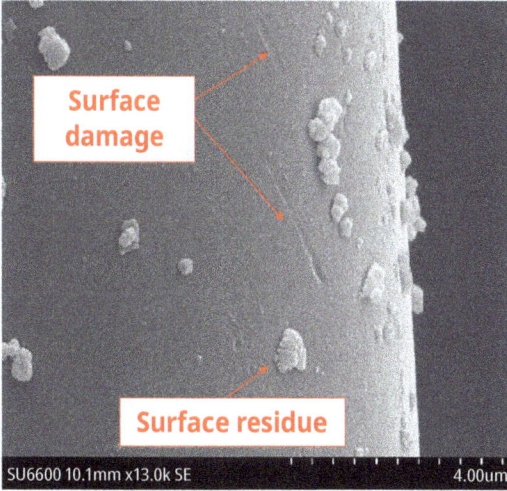

Figure 3.7: SEM images of the surface of rGF recycled from epoxy composites using the fluidised bed at 500 °C.

3.3.1.1 Combustion-fluidised bed recycling

Kennerley et al. [71] measured the strength of thermally recycled rGF using the fluidised bed process. Both recycling temperature and fluidisation velocity were varied, and the reported strength can be seen in Figure 3.8. Pickering et al. [10] observed similar strength loss when recycling using the fluidised bed process under the same conditions. Kennerley et al. [71] reported envisaging a reduction in rGF strength with fluidisation velocity, due to the additional agitation in the fluidised bed. No such relation was observed and the strength of rGF was not influenced by the fluidisation velocity.

The steady-state strength of GF thermally conditioned in [93] (Figure 3.6) is shown in Figure 3.8. The strength of recycled and conditioned fibres is comparable after exposure to 450 °C. However, rGF recycled in the fluidised bed at 550 and 650 °C are weaker than found in [93] at the same treatment temperature. Although it is not fully understood what causes the additional damage, it is most probably a result of mechanical damage induced by processing in the fluidised bed and/or thermal decomposition of the polymer matrix.

Thomason et al. [92] showed that rGF do not sustain more damage when recycled from GFRP in a furnace compared to thermally conditioning at the same temperature. On the contrary, Pender and Yang observed lower strength retention for rGF thermally recycled from GFRP statically in a furnace, when compared to thermally conditioned GF counterparts; suggesting that the additional processing and handling involved in retrieving fibres from the degraded GFRP may have resulted in additional fibre damage [78]. This suggests that the GF are likely mechanically damaged during processing within the fluidised bed recycling system. Additionally, Pender and Yang measured transient rise in

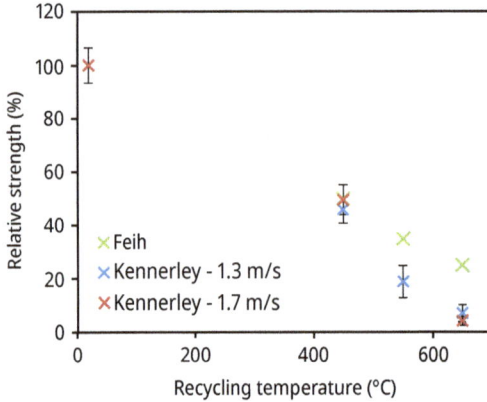

Figure 3.8: Comparison between tensile strength of rGF recycled using fluidised bed process and thermally conditioned GF, respectively (produced using data reported in [93] and [71]).

temperature (beyond set point recycling temperature) of >200 °C local to the GFRP during thermal recycling, which was attributed to rapid energy released during the highly exothermic combustion of the polymer matrix. Interestingly, the strength of GF exposed to analogous transient thermal conditioning time (<60 s) and temperature rise experienced during GFRP recycling shows a dramatic reduction in fibre strength of around 40% and 50% after exposure to 600 °C for merely 15 and 30 s, respectively. It was concluded that the brief increase in thermal loading caused by the polymer combustion can cause substantial weakening of the fibres [78].

Figure 3.9 shows the tensile strength of the rGF recycled using the catalysed combustion-fluidised bed, where the tensile strength of the virgin GF was found to be 2.55 ± 0.08 GPa. The recycling temperature does not appear to influence the tensile strength of the rGF. It is widely understood, however, that GF strength loss increases with the conditioning temperature [78]. The lack in such a trend in Figure 3.9 suggests that the mechanical damage subjected on the rGF, as they are processed through the fluidised bed recycling system, overwhelms any observable difference in thermal damage. Reducing the temperature required for the epoxy decomposition does not appear to have the added benefit of increasing the strength retention of fibres recycled in the fluidised bed used in this work. Pickering et al. reported an increase in fibre strength loss with fluidised bed temperature; recycling at 450, 550, and 650 °C yielded fibres with strength loss of 50%, 80%, and 90%, respectively [10]. Despite the similarities in the recycling process, the methods for fibre separation and recovery are entirely different. Pickering et al. [10] use a rotating sieve to separate the GF, which might reduce fibre abrasion and weakening. The cyclone separation method used by Pender et al. [10] may cause additional damage to the already thermally weakened GF.

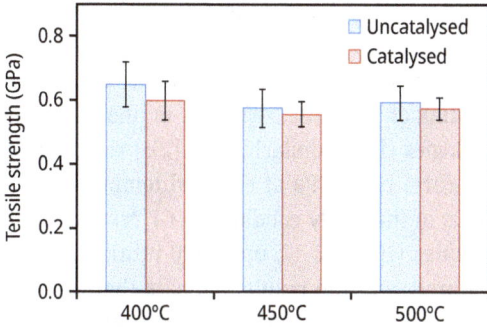

Figure 3.9: Effect of reactor temperature on tensile strength of rGF using catalysed combustion-fluidised bed (produced using data reported in [11, 66]).

3.3.1.2 Pyrolysis recycling

Williams et al. [39] observed a 45% drop in tensile strength after recycling GFRP using pyrolysis at 450 °C. Cunliffe et al. [99] also used pyrolysis for GFR recycling and the effect of recycling temperature was investigated. After liberating with pyrolysis, the rGF were cleaned of char residue by heating in air at the same temperature as was used for pyrolysis. Contrary to what is found when thermally conditioning GF, no clear correlation between the pyrolysis temperature and the tensile strength was found between 400 and 650 °C. No treatment time was given by Cunliffe et al. for the final thermal cleaning stage in air. The strength of thermally conditioned fibres is dependent on both temperature and time [93], hence inconsistent thermal cleaning time could explain the lack in correlation.

3.3.1.3 Microwave-assisted thermal recycling

Åkesson et al. [89] used microwave pyrolysis to recycle GF-UPR. GF bundles from a roving were infused with UPR and the strength of the bundle after thermal recycling was compared to that of the virgin-roving. Åkesson et al. [89] reported rGF bundles with strength loss of just 25% using microwave pyrolysis at 360 and 440 °C, again, no relation between pyrolysis temperature and fibre strength was observed. Since the change in bundle strength is given (opposed to single filament) this data cannot be directly compared to other GFRP pyrolysis studies [39, 99]; however, relative strength loss is typically higher when testing bundles after thermal weakening due to fibre-fibre abrasion during tensile testing [93].

Unlike strength, the tensile modulus of GF does not degrade following exposure to typical thermal recycling temperatures. In fact, several authors have observed a significant increase in GF modulus following heating above 400 °C. Otto [105] and

Yang and Thomason [106] observed GF length contraction and fibre modulus increased with conditioning temperature, which was attributed to entropy relaxation in the glass structure by Yang and Thomason. It was observed that the tensile modulus of GF could be increased by 14% following thermal loading of 500–550 °C which is in line with temperature used in thermal recycling technologies [106]. Similarly, Otto [105] reported GF modulus increase from 75 (room temperature) to 86 GPa at elevated temperature of 600 °C. Feih et al. reported that the modulus of thermally conditioned GF remained unchanged across a range of heating schedules (150–650 °C, up to 120 min); while no increase to the modulus of the GF was found, and the results indicate that thermal recycling will not degrade the stiffness of reclaimed GF [93]. Fraisse et al. [107] reported an rGF modulus of 87.5 GPa following combustion recycling statically in a furnace, which represented a 6.4% increase compared to virgin GF tested. Ginder and Ozcan [108] observed a 30% and 21% increase in pyrolysis-recycled rGF modulus from composite wind blade and SMC waste, respectively. Pickering et al. [10] observed no significant effect of heat on the modulus of rGF thermally recycled in the fluidised at temperatures of 450–650 °C, concluding that these results are encouraging given that the majority of the potential applications for these materials are compounded, short fibre composites which are stiff rather than strength-limited.

Table 3.3 presents the collated tensile strength of rGF recycled using a range of different thermal technologies and processing parameters. It is concluded that the strength of thermally recycled rGF is significantly lower than virgin GF counterparts, regardless of thermal process used. Strength retention ranges from 4% to 67% (for single filament testing) and is heavily dependent on the recycling temperature used. Pyrolysis recycling appears to produce rGF with a higher strength retention compared to the fluidised bed, mostly likely owing to additional mechanical damage induced during recycling in the fluidised bed.

Table 3.3: Collated tensile strength of thermally recycled rGF.

Recycling process	Waste type	Test type	Process parameter(s)	rGF strength		Ref.
				Unit	Retained	
Fluidised bed	Composite wind blade (GF-epoxy)	Single filament	500 °C	578 MPa	23%	[9]
Fluidised bed	GF-UPR	Single filament	450 °C	1,400 MPa	55%	[71, 98]
			550 °C	930 MPa	36%	
			650 °C	110 MPa	4.2%	

Table 3.3 (continued)

Recycling process	Waste type	Test type	Process parameter(s)	rGF strength		Ref.
				Unit	Retained	
Catalysed fluidised bed	GF-epoxy	Single filament	400 °C	601 MPa	24%	[11, 66]
			450 °C	570 MPa	22%	
			500 °C	570 MPa	22%	
Fluidised bed			400 °C	641 MPa	25%	
			450 °C	570 MPa	22%	
			500 °C	588 MPa	23%	
Pyrolysis	GF-UPR	Single filament	450 °C, 90 min	1,120 MPa	55%	[39]
Microwave-assisted pyrolysis	GF-UPR	Roving	360 °C	0.36 N/Tex	75%	[89]
			440 °C	0.35 N/Tex	73%	
Pyrolysis	GF-epoxy	Single filament	400 °C, 60 min	1,180 MPa	59%	[99]
			450 °C, 60 min	1,270 MPa	63%	
			500 °C, 60 min	1,080 MPa	53%	
			650 °C, 60 min	1,350 MPa	67%	
			400 °C, 60 min + 450 °C oxidation	1,060 MPa	52%	
			450 °C, 60 min + 450 °C oxidation	710 MPa	35%	
			500 °C, 60 min + 450 °C oxidation	1,060 MPa	52%	
			650 °C, 60 min + 450 °C oxidation	990 MPa	49%	
			800 °C, 60 min + 450 °C oxidation	360 MPa	18%	
Static combustion in furnace	GF-epoxy	Single filament	565 °C	400 MPa	20%	[107]

Table 3.3 (continued)

Recycling process	Waste type	Test type	Process parameter(s)	rGF strength		Ref.
				Unit	Retained	
Pyrolysis	SMC (GF-polyester vinyl ester blend)	Single filament	Up to 450 °C, 40 min	1,405 MPa	56%	[108]
	Composite wind blade (GF-epoxy)			1,311 MPa	60%	
Static combustion in furnace	GF-UPR	Single filament	500 °C, isothermal 25 min + 45 min heat up	902 MPa	34%	[109]
			550 °C, isothermal 25 min + 45 min heat up	718 MPa	27%	
			600 °C, isothermal 25 min + 45 min heat up	920 MPa	35%	
Static combustion in furnace	GF-epoxy	Single filament	500 °C, 80 min	940 MPa	37%	[66]

3.3.2 Thermally recycled carbon fibre

Exposing CF to typical composite thermal recycling temperatures has been demonstrated to significantly impact fibre tensile strength and modulus [110]. Feih and Mouritz [110] investigated the influence of thermal treatment conditions (temperature, time, and atmosphere) on CF. The mass loss of CF when heated in air or nitrogen was determined using TGA. Feih and Mouritz found that heating in both nitrogen and air caused a small loss in mass (0.8%) over the temperature range of 300–500 °C which was attributed to decomposition of the organic-based sizing compound on the CF. A large mass loss was, however, found when heated in air between 500 and 950 °C due to fibre oxidation. Prolonged heating in air eventually resulted in complete oxidation of the CF. A strong correlation was found between the mass loss and the diameter of CF when heated in air shown in Figure 3.10.

Feih and Mouritz also exposed CF bundles to temperatures between 250 and 700 °C, for up to 4 h, in both air and inert nitrogen atmosphere. Feih and Mouritz found that the fibre modulus remained unchanged when heated in nitrogen; however, the modulus decreased in air when fibres were heated above 550 °C as shown in Figure 3.11. These results suggest that the temperature alone is not responsible for the loss in stiffness which occurred in air. Feih and Mouritz concluded that the reduction in CF modulus in air is associated with the mass loss caused by surface oxidation.

Feih and Mouritz found that the reduction in CF strength was identical whether heating occurs in air or nitrogen. It was therefore concluded that the strength loss

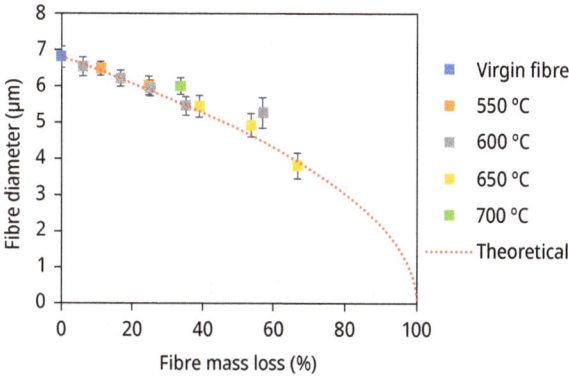

Figure 3.10: Relationship between CF mass loss and diameter following thermal conditioning in air at different temperatures and exposure times between 500 and 700 °C (produced from data reported in [110]).

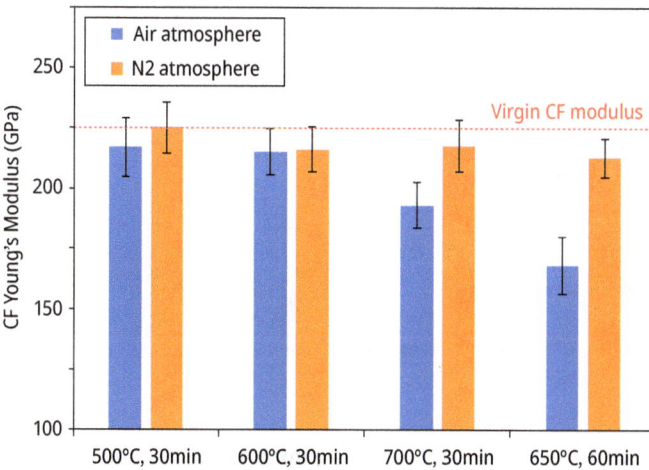

Figure 3.11: CF modulus following heat treatment in air or nitrogen (produced using data reported in [110]).

was dependent on the temperature but was not related to CF oxidation which occurred in air [110]. The effect of temperature on CF strength in air, as found by Feih and Mouritz, is shown in Figure 3.12 for two heat-treatment times of 30 and 120 min [110]. CF strength decrease onset occurs for temperatures greater than 400 °C, which is in line with typical composite thermal recycling temperatures. This work shows that the parameters used during thermal recycling of CFRP can have a significant impact on CF mechanical properties and should be carefully considered to yield an rCF product with optimal performance and value.

Feih and Mouritz proposed a phenomenological CF strength loss model (as shown in Figure 3.12) which can describe the strength of CF as a function of exposure temperature alone which is described in eq. (3.1).

CF tensile strength loss model is described by Feih and Mouritz in [110]:

$$\sigma(T) = \frac{\sigma_0 + \sigma_R}{2} - \frac{\sigma_0 - \sigma_R}{2} \tanh\big(k_f(T - T_{50\%})\big) \tag{3.1}$$

where $\sigma(T)$ is CF tensile strength (GPa) following exposure at temperature, T (°C), σ_0 is the room temperature CF tensile strength (GPa), σ_R is the steady-state-reduced CF tensile strength (GPa), $T_{50\%}$ is the temperature at which 50% of the fibre strength is lost (°C), and k_f describes the rate of CF tensile strength loss and is fitted as $k_f = 0.021$.

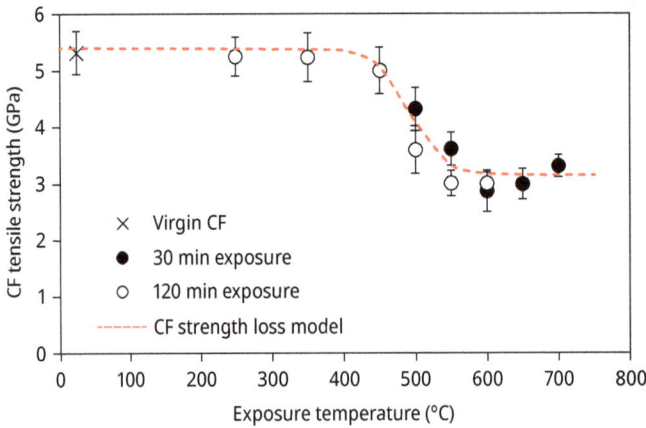

Figure 3.12: Effect of temperature on CF strength when heated in air for 30 and 120 min (produced using data reported in [110]).

Jiang and Pickering [111] studied the surface properties, microstructure, and mechanical properties of pyrolysed and oxidised rCF to gain insight into to the structure-property relationship for thermally recycled rCF. Waste prepregs were first pyrolysed at 500 °C, and then the char residue on the rCF surface was removed by oxidation in air. In agreement with Feih and Mouritz [110], Jiang and Pickering concluded that thermal recycling involves both thermal and oxidative effects [111]. Using XRD and Raman spectroscopy, Jiang and Pickering showed that the oxidative effect results in surface defects, and the surface defects causes a reduction in tensile strength and lateral crystallite size. The thermal effect of the recycling process results in an expansion in the distance between graphite layers and a decrease in surface oxygen concentration (measured using XPS). The tensile strength of rCF has a strong correlation with the intensity ratio of the D and G bands of the Raman spectra (ID/IG). With an increase in ID/IG, the tensile strength of rCF decreases linearly [111].

3.3.2.1 Combustion-fluidised bed recycling

Yip et al. [112] trialled the use of the fluidised bed recycling system (reactor temperature of 450 °C) for CF-epoxy components. Single fibre testing showed that recovered fibres retain ~ 75% of their tensile strength, while their Young's modulus remains unchanged. Property degradation of rCF was attributed to sand particle abrasion at high temperature during the fluidised bed process, the wall roughness of the cyclone separator, and fibre handling after the recycling process [112].

Pickering et al. [113] later used a pilot-scale-fluidised bed recycling pilot line to recover Toray T800 CF, commonly used in the aerospace industry, from a toughened epoxy end-of-life component. The study used a reactor temperature of 500 °C and the rCF had a mean tensile strength of 4.9 GPa, retaining 82% of the virgin fibre strength [113]. The mean elastic modulus of the rCF increased by 8% from the virgin CF equivalent [113]; however, the study did not consider this to be a statistically significant increase and therefore consistent with the earlier finding by Yip et al. [112], that the modulus is retained through the process rather than increased.

It remains unclear what the relative contributions of thermal, oxidative, and mechanical factors play in determining the strength degradation of fluidised bed-recycled rCF. Fluidised bed-recycled rCF generally exhibit a lower strength retention when compared to rCF thermally recycled using static processes at the same temperature [26, 31, 52], which suggests that sand attrition (and/or other mechanical forces) play an important role.

3.3.2.2 Pyrolysis recycling

The pyrolysis recycling method typically consists of thermal degradation of the matrix and a secondary oxidation step, which have different impacts on surface chemistry and mechanical properties of rCF [110]. It has been observed in several studies that temperature and time used during pyrolysis are key parameters in determining the quality of rCF [30, 52, 114].

Mazzocchetti et al. [25] demonstrated that secondary oxidation time is effective in determining the level of damage sustained by pyrolysis-recycled rCF. Mazzocchetti et al. [25] measured the effect of oxidation on both virgin CF and pyrolysis-recycled rCF, finding virgin CF has a higher propensity to damage during the oxidation process compared with pyrolysed rGF. Virgin CF experienced a 3% reduction in diameter following oxidation at 500 °C for 20 min, which was attributed by Mazzocchetti et al. to the removal of the sizing layer. Following extended oxidation for 60 min at 500 °C, fibres sustained a 10% diameter reduction due to the fibre decomposition. In contrast, oxidising pyrolysis-recycled rCF for 60 min at 500 °C resulted in comparable diameters with virgin CF. Mazzocchetti et al. theorised the presence of a char layer on the pyrolysed rCF, which resulted in 10% larger fibre diameter compared with the virgin

CF counterpart. These results suggest that the char layer played a protective role, limiting direct exposure of the rCF surface to the oxidising atmosphere, reducing the overall rate of fibre oxidation to CO and CO_2 [25]. Mazzocchetti et al. did not measure the mechanical properties of the fibres following oxidation (either virgin or pyrolysed); therefore this study alone does not have the power to establish a causal relation between level of fibre oxidation (measured as reduction in fibre diameter) and the residual fibre properties.

Yang et al. [29] attempted to improve the efficiency of pyrolysis recycling by controlling oxygen gas concentration in the atmosphere during pyrolysis. The role of oxygen concentration, pyrolysis temperature, and duration was studied with respect to rate of degradation and mechanical properties of rCF. Pyrolysis of CF-epoxy was carried out at a range of oxygen concentration (5, 10, and 20 vol.% with the remaining atmosphere comprising nitrogen). Temperatures of 550, 600, and 650 °C were investigated for durations ranging 15–60 min. Yang et al. observed mass loss increased with oxygen concentration, temperature, and pyrolysis duration. This was attributed to greater removal of pyrolytic carbon char layer in addition to oxidation of the rCF themselves [29]. It was found that increasing oxygen content from 5 to 10 vol.% had the largest influence on residual fibre tensile strength, as is shown Figure 3.13. Further increasing oxygen concentration to 20 vol.% had no significant impact on rCF strength; however, a decline in the tensile modulus was observed to initiate under these conditions [29]. Thermal processing at 650 °C in the presence of 5 vol.% oxygen for 45 min was found to be the optimum condition analysed by Yang et al. [29], which yielded 80% retention in rCF strength. The work conducted by Yang et al. demonstrated that balancing the temperature, duration, and gas composition during pyrolysis is critical to maintain rCF performance.

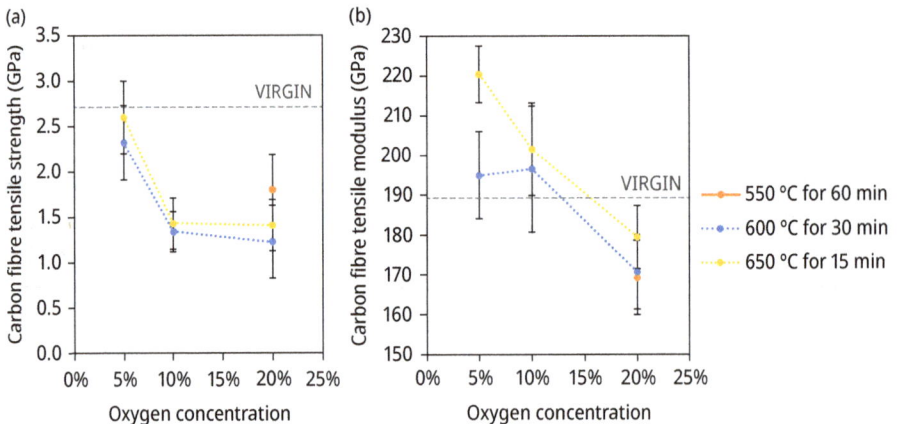

Figure 3.13: Effect of oxygen concentration and temperature during pyrolysis on rCF: (a) tensile strength and (b) modulus (graph produced using data from [29]).

Nahil and Williams [52] investigated the influence of pyrolysis temperature on CF-polybenzoxazine composites for 1 h at 350, 400, 450, 500, and 700 °C. A post-oxidation process was then carried out at 500 and 700 °C to remove the pyrolytic char residue. Nahil and Williams [52] observed a maximum retention of rCF mechanical properties when using a pyrolysis and oxidation temperature of 500 °C, where 93% and 96% of the tensile strength and modulus were retained, respectively. It was found that the duration of the secondary oxidation step could be reduced from 2 h to just 15 min by increase oxidation temperature from 500 to 700 °C. Despite the reduced exposure duration, Nahil and Williams [52] found that rCF oxidised at 700 °C suffered a 64% reduction in tensile strength. This is significantly greater than was found for 500 °C oxidation, suggesting that processing rates likely need to be limited to avoid excessive rCF weakening.

Meyer et al. [26] investigated the influence of thermal recycling parameters on the reclamation of rCF from CF-epoxy. Temperatures between 400 and 900 °C in both air and nitrogen atmospheres were used with the aim of removing the epoxy matrix and evaluating the integrity of the rCF. In an air-based atmosphere, the oxidation of rCF at temperatures of above 600 °C caused greater mass loss than that of the original epoxy content (see Figure 3.14) [26]. It was proposed that oxidation of rCF could result in a decrease of the tensile strength and Meyer et al. [26] concluded that pyrolysis in the presence of oxygen should be avoided at temperatures above 600 °C. This is in contradiction with the findings from Feih and Mouritz [110] and was not substantiated with data showing the impact of pyrolysis atmosphere on fibre strength [26]. When using a nitrogen atmosphere, it was found that pyrolysis limits the oxidation of the rCF, as indicated by the retained mass following pyrolysis approximating the fibre mass in the pyrolysed composite samples (see Figure 3.14). The study found that the full mass of the epoxy matrix was removed from the fibres at 550 °C and above in a nitrogen atmosphere [26]. An optimised pyrolysis cycle was proposed: a first stage in nitrogen at 550 °C for 2 h to remove the epoxy matrix; cool down to 200 °C; and an oxidation stage at 550 °C to remove the pyrolytic carbon. Meyer et al. [26] then measured the single fibre tensile strength for the optimised pyrolysis cycle, reporting a mean value of 3.6 GPa for the rCF, retaining 96% of the original strength. Meyer et al. [26] concluded that their findings indicate that the rCF is not damaged by oxidation using the optimised pyrolysis cycle.

Onwudili et al. [30] used pyrolysis to recycle rCF from CF-polybenzoxazine waste. A two-stage thermal process was employed: initially pyrolysing in nitrogen at 400 °C, followed by an oxidation stage at 450 °C for 2 h to remove char residues [30]. The study reported tensile strength results for rCF following isothermal dwell times of 0, 30, and 60 min during the pyrolysis stage at 400 °C. The retained tensile strengths of the rCF were 93%, 78%, and 77% fibre following isothermal dwell times of 0, 30, and 60 min, respectively [30]. Both the tensile strength and modulus values decreased with longer isothermal dwell times. It remains unclear whether the strength loss in rCF observed by Onwudili et al. [30] is a result of the initial pyrolysis step or the secondary (significantly longer duration) oxidation step at 450 °C.

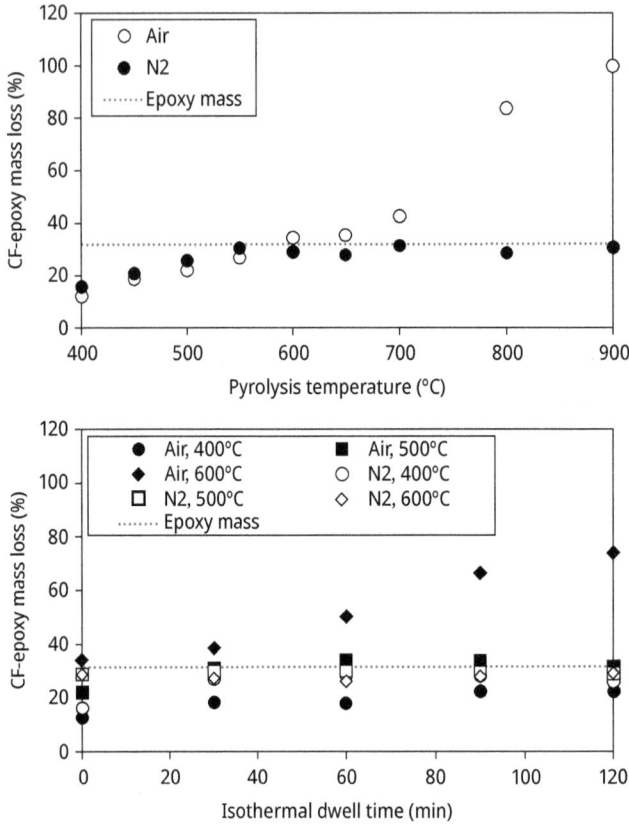

Figure 3.14: Influence of pyrolysis parameters on epoxy matrix removal and rCF degradation (produced using data reported in [26]).

Kim et al. [31, 34, 115] have performed several parametric studies investigating steam pyrolysis recycling of CF-epoxy composites. As with other studies, Kim et al. [31] employed a two-step process, starting with initial pyrolysis followed by heating the reclaimed fibres in air to remove surface residue. Kim et al. [31] conducted several temperature and time scenarios, observing optimal rCF strength retention of 90% when conducting pyrolysis (30 min) and secondary oxidation (60 min) at 550 °C. There was a general trend that strength reduced with increasing isothermal dwell time during the air cycle [31]. Though less significant, the same trend was observed with the elastic modulus of rCF. The rCF exposed to an air cycle of 60 min retained 90% of the virgin fibre modulus [31].

In a separate study, Kim et al. [115] investigated the use of steam pyrolysis only (excluding secondary oxidation step), with varying temperatures between 600 and 800 °C and treatment time of 1 h. The retained strength values of rCF were considerably lower than those in their other study at 550 °C using both steam and air atmosphere cycles, with a maximum mean strength retention of 66% observed at 800 °C [115]. Similar strength retention was observed at the minimum temperature cycle of 600 °C; however, pyrolysis at temperatures of 650, 700, and 750 °C yielded significantly lower rCF strengths. The lowest mean strength retention was 38% after recycling at 700 °C. The retained modulus, however, was above 80% for all experiments in the study and effectively retained at 800 °C with a retention of 101% [115]. The results suggest that pyrolysis in steam atmospheres requires additional cycles in air atmosphere to achieve the best strength retention.

Ye et al. [35] conducted pilot-scale pyrolysis of CF-epoxy composites using a mixed atmosphere of steam and nitrogen. Ye et al. observed a mean strength retention of 83% when pyrolysing at 600 °C, which is higher than was found by Kim et al. [115] using steam only. The higher strength retention compared to that found by Kim et al. suggests the inclusion of nitrogen during steam pyrolysis may limit strength degradation and yield advantages to resulting rCF performance.

Another study by Kim et al. [34] used a two-stage process of CO_2 pyrolysis at 400 °C followed by steam pyrolysis at 700 °C to recycle CF-epoxy composites. The isothermal dwell time was varied during the steam stage, and it was found that the strength retention of rCF was higher for dwell times of 20 or 40 min, 86% and 90%, respectively, and less than 80% for dwell times of 60, 80, and 100 min [34]. The retained modulus was higher than 98% for all variants of isothermal dwell time. For this methodology, rCF strength retention is shown to be dependent on isothermal dwell time, whereas fibre modulus is unaffected by recycled time (across durations investigated) [34].

3.3.2.3 Microwave-assisted thermal recycling

An early feasibility study by Lester et al. [87] measured the mechanical properties of rCF recycled from CF-epoxy laminates using microwave heating in nitrogen. Tensile strength and modulus retention were found to be 72% and 88%, respectively. The temperature used during recycling was not reported making it challenging to compare the impact of microwave heating against other conventional methods [87].

Two recent studies have recovered rCF from CF-epoxy composites, both with microwave frequencies of 2.45 GHz [2, 114]. Firstly, Hao et al. [114] used a cycle of microwave pyrolysis lasting 15 min with maximum microwave power of 3 kW and trialled temperatures of 450, 550, and 650 °C. An oxidation step of 550 °C for 30 min is followed. The retained tensile strengths of the rCF were 87%, 83%, and 81% for the respective temperatures, with strength reducing with increasing temperature [114].

Ren et al. [2] also observe an increased tensile strength with increasing temperature. The study, however, only trialled pyrolysis temperatures of 400 and 500 °C for 15 min and the microwave used had a maximum power output of 1 kW. Different temperatures and isothermal dwell times were also investigated for the subsequent oxidation step and the optimal conditions found to be 550 °C for 30 min for retaining strength [2]. The elastic modulus, however, was higher for an oxidation step of 550 °C for 40 min. Ren et al. [2] report a full retention of rCF strength with a pyrolysis temperature of 500 °C and an increase in modulus compared to virgin CF.

Table 3.4 summarises the collated mechanical properties of thermally recycled rCF. It is concluded that the strength retention of thermally recycled rCF is general found to decrease with increasing recycling temperature, thermal exposure time, and oxygen content present in the process. Pyrolysing in nitrogen produced rCF with strength retention range from 36% to 99%, with strength retention heavily dependent on both pyrolysis and secondary oxidation temperature. Pyrolysis recycling temperatures should be limited to 550–650 °C to minimise rCF-weakening; however, higher temperatures in steam atmospheres are conducive to improved rCF strength retention. Recycling using the fluidised bed results in greater strength for rCF compared to pyrolysis when similar recycling temperatures are used. Regardless of the thermal recycling used, rCF modulus is well maintained and often higher than the virgin CF counterpart.

Table 3.4: Collated mechanical properties of thermally recycled rCF.

Recycling process	Materials	Process parameters	rCF strength		rCF modulus		Ref.
			GPa	Retained	GPa	Retained	
Fluidised bed	CF-epoxy	450 °C	2.85	73%	227	99%	[112]
Fluidised bed	CF-epoxy	500 °C	4.90	72%	334	108%	[113]
Pyrolysis	CF-epoxy	550 °C, 60 min, air	1.8	66%	169	89%	[29]
		600 °C, 30 min, air	1.2	45%	171	90%	
		600 °C, 30 min, 10% O_2/90% N_2	1.3	49%	197	104%	
		600 °C, 30 min, 5% O_2/95% N_2	2.3	86%	195	103%	
		650 °C, 15 min, air	1.4	52%	179	95%	
		650 °C, 15 min, 10% O_2/90% N_2	1.4	53%	202	107%	
		650 °C, 15 min, 5% O_2/95% N_2	2.6	96%	220	116%	
		650 °C, 30 min, 5% O_2/95% N_2	2.2	81%	196	103%	
		650 °C, 45 min, 5% O_2/95% N_2	2.1	78%	155	82%	

Table 3.4 (continued)

Recycling process	Materials	Process parameters	rCF strength		rCF modulus		Ref.
			GPa	Retained	GPa	Retained	
N₂ pyrolysis	CF-epoxy	550 °C, 2 h +550 °C oxidation	3.57	96%	N/A	N/A	[26]
N₂ pyrolysis	CF-polybenzoxazine	400 °C, 0 min +450 °C oxidation	N/A	93%	N/A	128%	[30]
		400 °C, 30 min + 450 °C oxidation		78%	N/A	124%	
		400 °C, 60 min + 450 °C oxidation		77%	N/A	104%	
N₂ pyrolysis	CF-polybenzoxazine	350 °C, 1 h +500 °C oxidation	2.3	67%	235	98%	[52]
		400 °C, 1 h +500 °C oxidation	2.8	80%	235	98%	
		450 °C, 1 h +500 °C oxidation	2.9	82%	233	97%	
		500 °C, 1 h +500 °C oxidation	3.3	93%	230	96%	
		700 °C, 1 h +500 °C oxidation	2.7	78%	230	96%	
		350 °C, 1 h +700 °C oxidation	0.9	26%	182	76%	
		400 °C, 1 h +700 °C oxidation	1.1	30%	233	97%	
		450 °C, 1 h +700 °C oxidation	1.2	33%	230	96%	
		500 °C, 1 h +500 °C oxidation	1.3	36%	223	93%	
		700 °C, 1 h +700 °C oxidation	1.2	35%	223	93%	
Steam pyrolysis	CF-epoxy	550 °C steam, 30 min 550 °C air, 1 h	4.26	83%	193	95%	[31]
		550 °C steam, 45 min 550 °C air, 1 h	3.53	79%	200	99%	
		550 °C steam, 60 min 550 °C air, 1 h	3.35	90%	198	98%	

Table 3.4 (continued)

Recycling process	Materials	Process parameters	rCF strength		rCF modulus		Ref.
			GPa	Retained	GPa	Retained	
		550 °C steam, 75 min 550 °C air, 1 h	3.85	79%	198	98%	
		550 °C steam, 90 min 550 °C air, 1 h	3.37	75%	195	96%	
Steam pyrolysis	CF-epoxy	700 °C, 1 h	1.53	38%	159	81%	[115]
		800 °C, 1 h	2.68	66%	197	101%	
Steam pyrolysis	CF-epoxy	600 °C	3.64	83%	N/A	N/A	[35]
N_2 pyrolysis + steam pyrolysis	CF-epoxy CF-epoxy	400 °C CO_2 700 °C steam, 20 min	3.41	86%	206	98%	[34]
		400 °C CO_2 700 °C steam, 40 min	3.55	90%	206	98%	
		400 °C CO_2 700 °C steam, 60 min	3.05	77%	211	100%	
		400 °C CO_2 700 °C steam, 80 min	2.94	74%	210	99%	
		400 °C CO_2 700 °C steam, 100 min	3.11	79%	216	102%	
Microwave pyrolysis	CF-epoxy	450 °C, 15 min +550 °C oxidation	4.08	87%	201	95%	[114]
		550 °C, 15 min +550 °C oxidation	3.87	83%	204	96%	
		650 °C, 15 min +550 °C oxidation	3.77	81%	197	93%	
Microwave pyrolysis	CF-epoxy	400 °C, 15 min +550 °C oxidation	2.72	89%	237	107%	[2]
		500 °C, 15 min +550 °C oxidation	3.06	99%	246	111%	

3.4 Fibre-polymer adhesion of recycled fibre

3.4.1 Influence of thermal recycling temperature of fibre sizing

Sizings are ubiquitously applied to the surface of reinforcement fibre during production and are integral to ensuring that the desired level of fibre-polymer adhesion is achieved in FRP. Although the exact composition of sizings are typically confidential and vary depending on the polymer matrix material, most contain an organic film former and an organofunctional silane coupling agent [116]. Several authors have studied the thermal behaviour of sizings [117–119] and organosilanes [94, 120], using TGA in the temperature ranges associated with composite thermal recycling. Nagel et al. [117] performed TGA on PP compatible commercially sized GF in air and found that most of the sizing is degraded below 300 °C. Both Nagel et al. [117] and Feih et al. [93] observed a reduction in sizing thermal stability when heated in air opposed to nitrogen. Gao and Su [119] thermally degraded two different commercial sizings using TGA, observing that both sizings were degraded at around 320 °C in air. Rudzinsk et al. [118] performed TGA on aminopropyltriethoxysilane (APS) coupling agent in combination with several commonly used films. It was found that all the sizings fully degraded by a temperature of 550 °C when pyrolysed in nitrogen, with peak rates of decomposition occurred at around 350 °C. Jenkins et al. studied the thermal behaviour of an APS film using simultaneous TGA and DSC under air. It was observed that the APS decomposition completed at 650 °C, with the organics starting to degrade around 300 °C [94]. Pham and Chern [120] studied the thermal decomposition of four other organosilanes in addition to APS. Although they showed different thermal behaviour, all silanes decomposed between 300 and 650 °C [120], similar to the observations of Jenkins et al. [94]. The peak rates of mass loss occurred between 400 and 550 °C [120]. Based on the thermal analysis data it is clear that exposure to typical thermal recycling temperatures results in the removal of the sizing which can ultimately reduce recycled fibre compatibility with polymer systems when reused as composite reinforcement elements [94, 117–120].

3.4.2 Glass fibre-polymer adhesion

Using thermally recycled or thermally conditioned GF as reinforcement produces composites with diminished mechanical performance due to the poor fibre strength and interfacial adhesion between the GF and polymer matrix [32, 42, 89, 91, 92]. Several authors have used micromechanical techniques to measure the interfacial shear strength (IFSS) between thermally conditioned GF and a PP matrix [91, 92, 117, 121]. Yang et al. [91] and Thomason et al. [92] thermally conditioned GF at 500 °C for 25 min in air. Using the microbond test, Yang et al. [91] reported a drop in apparent IFSS from 16 to 8 MPa after thermal conditioning. In agreement, Thomason et al. [92] observed a reduction in apparent IFSS from 14.7 to 9.2 MPa. Nagel et al. [117] also used the microbond test to investigate the effect of thermal conditioning temperature and atmosphere on apparent IFSS between GF and PP. Nagel et al. observed a correlation between the temperature at which the apparent IFSS reduced significantly (250 °C in air) and temperature when sizing decomposed (between 200 and 250 °C as measured using TGA); they ultimately concluded that thermal decomposition of the sizing resulted in reduced adhesion with PP, lowering the measured IFSS [117].

Pender [66] used the microbond test to measure the IFSS between rGF (recycled from GF-epoxy in the fluidised bed) and both polypropylene (PP) and epoxy resin. The adhesion in each scenario was compared to virgin GF coated with APS alone. It was concluded that IFSS with PP was unchanged between the virgin and rGF; however, the IFSS in each case was lower than was observed on other studies. On the contrary, IFSS between rGF and epoxy was reduced by approximately 40% compared to virgin GF, most likely due to APS degradation. This was confirmed by observed rebound in IFSS with epoxy when APS was reapplied to the rGF [66]. In addition to micromechanical techniques, Bikiaris et al. [122] and Åkesson et al. [89] used fracture surface analysis to conclude that the interfacial adhesion is diminished after thermally conditioning GF at typical composite recycling temperatures.

3.4.3 Carbon fibre-polymer adhesion

Early research conducted by Zielke et al. [123] using XPS suggests that oxygen functionalities on CF surface can be removed in the form of CO and CO_2 by heating at high temperatures in an inert atmosphere. This would suggest a reduction in rCF surface functionality following pyrolysis recycling, potentially limiting interaction and adhesion with new resin systems.

There remains no clear consensus in the literature as to whether the impact of thermal recycling is positive [31, 124], negative [111], or neutral [111, 125, 126], with respect to the adhesion between rCF and polymers. It is proposed that secondary oxidation treatments of rCF (as in the case of secondary oxidation of pyrolysed rCF) can increase oxidised groups on the rCF surface which can act as cross-linkers with new resins. Several authors report the O/C ratio on the rCF surface following recycling, using this as a proxy for surface functionality and, by extension, affinity to bond with new resins [25, 29, 111]. The atomic O/C ratio is a quantitative measurement of oxygen containing functional groups on the CF surface and is a good indicator of the amount of active surface area of the CF available for chemical bonding with resin.

Mazzocchetti et al. [25] used EDX to measure the oxygen concentration on pyrolysed rCF following secondary oxidation treatments. Oxygen concentration was found to increase by 7% and 72% after oxidation at 500 °C for 20 and 60 min, respectively (compared to pyrolysed rCF without secondary oxidation) [25]. These finding are corroborated by Wang et al. who showed that the concentration of oxygen gas was the dominant factor in determining the oxygen content on fibres surface following pyrolysis recycling; where increasing atmospheric oxygen content produced higher O/C ratio on the fibre surface [29]. In either case, IFSS following oxidation was not directly measured; therefore it is unclear how this impacts adhesion with new resin systems [25, 29].

Jiang and Pickering [111] studied the surface properties of pyrolysed and oxidised rCF using XPS, Raman spectroscopy, and XRD. Waste prepregs were first pyrolysed at 500 °C and then the char residue on the rCF surface was removed by oxidation in air. In agreement with Mazzocchetti et al. [25] and Wang et al. [29], the oxidation process was found to introduce new oxidised groups on the rCF surface. Despite this, it was observed that the combined pyrolysis and oxidation process still reduced overall oxygen content on fibres compared to virgin CF. They reported that O/C ratios of virgin and recycled CF were 0.26 and 0.15, respectively, based on XPS analysis. This is in close alignment with Fernández et al. [127] who observed an O/C of 0.18 for commercially pyrolysed rCF. Similarly, Yip et al. [112] observed a 16% reduction in O/C ratio following recycling of CF-epoxy composites using the fluidised bed when compared to virgin CF.

Jiang and Pickering [111] compared the IFSS (measured using single-fibre pull-out) of virgin and recycled GF with epoxy and PP. IFSS with epoxy was found to reduce by approximately 25% following recycling, whereas no change in IFSS with PP was observed [111]. This reduction in IFSS may be attributable to the observed reduction in O/C ratio; however, further investigation would be needed to establish this as a causal mechanism.

Jiang et al. [125] studied the surface properties of rCF recycled from epoxy composites, which were compared to virgin CF after sizing removal. The exposure of rCF to hot oxidative atmosphere during recycling (fluidised bed, 550 °C) was found to convert some surface hydroxyl groups into higher oxidation state (carbonyl ($C=O$) and carboxylic groups (–COOH)) while maintaining a consistent O/C ratio. This change in surface chemistry did not influence the measured IFSS (using microbond test) of the rCF with epoxy resin. It was concluded that recycling rCF through thermal oxidative decomposition of polymer matrix did not weaken the interfacial bonding properties compared to de-sized virgin CF [125].

Kim et al. [31] measured IFSS (using the microbond test) between epoxy and rCF that had been reclaimed through their steam pyrolysis process; observing an improvement compared to de-sized virgin CF. The de-sized virgin CF had an average IFSS of 39 MPa, whereas the rCF had an average value of 47 MPa [31]. With increased isothermal dwell time in the steam pyrolysis process, greater oxygen content was observed on the rCF surfaces using FTIR, through the introduction of –OH, –CH–, H–bond, $–C=O$, and C–O functional groups. This resulted in a more polar rCF surface and an increase in surface-free energy calculated using dynamic contact angle measurement.

Wada et al. [124] also found an improved IFSS between epoxy and rCF compared to the unsized virgin fibre using the single fibre fragmentation test. The rCF were reclaimed using an induction-heating steam system, with isothermal dwell times of 5 min between 500 and 800 °C. rCF that were reclaimed at temperatures of 700 and 800 °C had higher IFSS values than the sized virgin fibre when a 4% volume content of nitrogen gas was included in the steam treatment.

Other studies have investigated methods for improving interfacial bonding between thermally recycled rCF and new resin systems [126–131]. Cai et al. [126] studied the interfacial adhesion between PP and rCF recycled from PA6 sheets using steam pyrolysis (650 °C for 1 h). Cai et al. compared the IFSS of virgin and recycled CF, measured using the microdroplet test. The results indicated that the IFSS of fibres recycled using steam pyrolysis approximated those of virgin fibre counterparts. SEM observation indicated that recycling did not induce changes in fibre surface morphology despite the effective removal of the matrix resin. Cai et al. [126] used XPS to analyse the fibre surface state, concluding that the fibre sizing was removed during recycling. In agreement with Kim et al. [31] and Wada et al. [124], Cai et al. observed the introduction of oxygen-containing groups on the fibre surface during steam pyrolysis, attributed to surface oxidation of rCF. Cai et al. concluded that chemical interactions such as hydrogen bonding between functional groups and the matrix resin resulted in the desirable interfacial adhesion properties of the recycled fibres, which could be further enhanced by the use of 0.5 wt.% maleic anhydride-grafted PP (MAPP) [126]. Similar observations were made by Wong et al. [130], who observed approximately a three-fold increase in IFSS of fluidised bed-recycled rCF with the use of MAPP coupling agent at 2 wt.% compared to PP homopolymer.

Burn et al. [131] investigated the surface morphology and functionality of pyrolysed CF surface and measured IFSS with PP. A pseudo-recycling method was used, where virgin CF tows were pyrolysed at 550 °C for 10 min to mimic typical pyrolysis recycling from CFRP waste. Burn et al. found that removal of the sizing from the CF using pyrolysis altered the surface morphology of the fibre. Using AFM it was found that RMS roughness of pyrolysed fibres reduced by 62% (compared to virgin), whereas fibre surface O/C ratio reduced by almost 50% (measured using XPS) [131]. This agrees with Lester et al. [87], who observed a reduction in fibre surface roughness (also measured using AFM) following microwave-assisted pyrolysis recycling from epoxy-based composites. Despite these findings, no significant reduction in IFSS was measured with PP matrix following pyrolysis when using the microdroplet test. In agreement with Cai et al. [126] and Wong et al. [130], a significant improvement in the fibre-matrix adhesion was achieved by adding a MAPP at 2 wt.%, which increased the IFSS by 330% compared to PP homopolymer [131]. Given that CF were not recovered from a polymer matrix [131], the potential interaction(s) of polymer thermal degradation on rCF surface morphology and functionality were not considered.

In conclusion, there is clear consensus across the literature that sizings used for both CFRP and GFRP thermally decompose at typical FRP recycling temperatures. The removal of surface sizing on thermally recycled rGF is attributed to the observed reduction in IFSS of 40–50% (with PP and epoxy) compared to virgin-sized GF. There remains no clear consensus in the literature as to whether the impact of thermal recycling is positive, negative, or neutral with respect to the adhesion between rCF and polymers. Surface functionality of rCF is typically measured in O/C ratio, with higher ratios indicative of greater number of functional groups to react with new resin systems. Many authors have used secondary oxidation treatment on rCF, with most seeing an increase in O/C ratio (however typically lower than virgin CF counterparts). To date, no study has systematically evaluated the relationship between O/C ratio and IFSS of virgin CF, and it remains unclear how important oxygenation of the rCF fibre surface for improving surface functionality.

3.5 Performance of composites reinforced with thermally recycled fibres

3.5.1 Remanufacturing composites using thermally recycled fibres

Thermally recycled fibres have historically proven challenging to integrate into composite manufacturing due to the low-density, randomly oriented, discontinuous fibre format that results from typical thermal recycling. Several limitations restrict the processing methods that can be used for manufacturing FRP with thermally recycled fibre. For ease of transport and efficient reactor loading, scrap FRP often requires size reduction prior to recycling. Thermally recycled fibres therefore have a distribution in fibre length; meaning they can only be used as discontinuous reinforcement in FRP. The sizing, which binds new fibres together, is also decomposed during recycling which allows the fibres to move and separate during processing. The fibre strand integrity and alignment are therefore lost, and fibres become entangled and fluffy with low bulk density.

Several authors have processed thermally recycled rGF into GFRP [4, 42, 66, 71, 89, 132]. Pender [66] used a papermaking process to mix fluidised bed-recycled rGF with PP fibres to produce glass mat thermoplastic, which was subsequently compression-moulded to form GF-PP plates. Åkesson et al. [89] produced randomly orientated mats with rGF, which were then infused with UPR. Li et al. [42] extruded rGF with PP into threads using a screw extruder which was then injection-moulded. Most of the studies, however, produced a BMC by dispersing the rGF in UPR [4, 71, 132]. Other authors have processed thermally conditioned GF into discontinuous GFRP using methods that may be suitable for thermally recycled rGF [92, 121, 133]. Roux et al. also used a twin-screw extruder to process PP with thermally conditioned GF. Thomason et al. [92] and Lee and Jang [133] both used papermaking processes to make GF-PP using thermally conditioned GF; Lee and Jang [133] were unique in their use of powdered PP in the dispersion (instead of co-mingling with chopped PP fibres). It has therefore been demonstrated that thermally recycled rGF are suitable to be used in discontinuous, randomly orientated GFRP using extrusion, papermaking, and BMC processes.

Several authors have processed thermally recycled rCF into CFRP. Three distinct classes of intermediate materials have been produced: moulding compounds, non-woven mats, and aligned tapes/prepregs. Pickering et al. [134] produced BMC using fluidised bed-recycled rCF, which were compression-moulded to produce the final composite material. Other studies have compounded pyrolysis-recycled rCF with thermoplastics (PP [127, 130], nylon 6 [135], polylactic acid (PLA) [136], polycarbonate (PC)/ABS copolymer [137], poly(butylene succinate) (PBS) [138], polyphenylene sulphide [55], poly(butylene terephthalate) [139]) which were then used as feedstock material in injection-moulding.

An array of dry [140] and wet [141–145] methods has been investigated to produce non-woven mat intermediates using thermally recycled rCF. rCF can be co-mingled with thermoplastic materials during the non-woven production which are then consolidated into composite sheets using compression-moulding [140, 144]. An alternative method was demonstrated by Giannadakis et al. [141] and Szpieg et al [143], who produced composite sheets by compression-moulding layers of pyrolysis-recycled rCF non-woven mats sandwiched between layers of PP films. Other authors have used RTM to infuse pre-prepared rCF non-woven mats with a liquid epoxy resin systems in production of CFRP [142].

Improved mechanical properties can be achieved when rCF are aligned in CFRP. The optimal mechanical properties of CFRP are typically attained with fibre volume fractions approaching 60%, which can only be achieved using aligned fibres. Several authors have investigated methods for re-aligning short-recycled fibres to improve the properties of composites reinforced using rCF [146, 147]. Numerous alignment methods have been proposed; however, "wet" fibre alignment methods is the most common and technically advanced approach utilised to date; where the alignment is achieved by forcing suspended rCF to follow the fluid streamlines though nozzles or alignment channels. Aligned rCF preforms can be deposited on resin films to produce tapes/prepregs or liquid resin infused.

Several studies have thermally conditioned/recycled CF [148] and GF [93, 98], statically without downsizing or disturbing the original fibre fabric architecture. These have then been reinfused with liquid resin to produce composite panels for mechanical testing. Although suitable for isolating and investigating solely the effect of fibre property and fibre-polymer adhesion change on FRP performance, these composites will most likely not comply with the architecture of commercial FRP based on thermally recycled discontinuous fibres.

3.5.2 GFRP reinforced with recycled glass fibres

A number of authors have reported the mechanical properties of GRFP reinforced with either thermally conditioned GF [91, 92, 98, 121] or thermally recycled rGF [4, 66, 71, 89, 132]. The tensile and flexural strength of GRP based on thermally recycled rGF tend to be inferior to those made with virgin fibres, irrespective of the thermal recycling method used, as seen in Table 3.5. Many authors attempt to mitigate this performance drop by reinforcing with both recycled and virgin GF. "Virgin fibre content" in

Table 3.5 gives the weight percentage of virgin GF used relative to total GF weight. Cunliffe and Williams [4] and Marco et al. [132] report high strength retention of GFRP; however, only 25% and 15% of the GF content is replaced with rGF. More notably, Kennerley et al. observed no change on GFRP strength when replacing up to 50% of GF content with rGF recycled using the fluidised bed. The drop in composite properties observed in Table 3.5 is likely a result of strength loss of the rGF and/or reduced adhesion between the fibres and the matrix. Feih et al. [98] established that the tensile strength of continuous GF-reinforced vinyl ester is solely controlled by the GF strength, with the relative strength loss of thermally conditioned GF mapping that of composites based on those fibres. In contrast, Nagel et al. [117] attributed the drop in tensile strength of injection-moulded PP composites to the loss of adhesion, when using thermally conditioned GF. It is understood that the IFSS has a larger effect on the strength of composite reinforced with shorter fibres [149], which could explain the conflicting observations made by Feih et al. [98] and Nagel et al. [117].

The modulus of discontinuous GFRP depends mainly on the fibre content [150–153] and fibre orientation [154–156], in addition to the polymer and GF moduli. With most rGF being used in discontinuous applications, use as an additive to increase polymer stiffness is another route to market for rGF, which may be ideally suited to applications in sectors such as automotive. Several studies have reported the modulus of GFRP prepared with rGF. Kennerley et al. [71] observed no change to BMC tensile or flexural stiffness when replacing all virgin GF with fluidised bed-recycled rGF. On the contrary, Cunliffe et al. [4] reported on average a 19% reduction in flexural modulus of BMC when replacing just 25 wt.% virgin GF with pyrolysis-recycled rGF in the compound mix. A large spread in data was observed; however, it means significance in the data was not established. Li et al. [42] were able to increase the flexural modulus of neat PP by 68% with the addition of 30 wt.% pyrolysis-recycled rGF; no virgin GF counterpart was measured for comparison. Pender [66] observed a 17% reduction in tensile modulus of PP composites (fibre fraction of 30 wt.%) when replacing all virgin GF with fluidised bed-recycled rGF. Pender [66] also observed a 96% increase in tensile modulus of neat epoxy with the addition of 20 wt.% fluidised bed-recycled rGF; no virgin GF counterpart was measured for comparison. These studies show that there could be more value in using rGF in applications where stiffness is required, opposed to load-carrying capacity, given the significantly lower knockdown in composites modulus (compared to strength data in Table 3.5) when using rGF.

Several authors have studied the use of polymer modification [66, 157], and/or sizings/coupling agents [66, 71, 157, 158], in the preparation of GFRP with thermally conditioned GF and thermally recycled rGF. Nagel [157] investigated the effect of MAPP content on injection-moulded PP prepared using thermally conditioned GF. Tensile strength was found to increase from 36 to 44 MPa with the addition of 4 wt.% MAPP. Treating the thermally conditioned GF in APS prior to moulding further increased the composite strength when used in connection with MAPP-modified PP matrix. Over 50% of the composite strength loss was recovered with APS treatment and just 1 wt.%

MAPP content. It was concluded that treatment of the fibres with APS is more effective in improving the tensile strength than solely the addition of MAPP to the matrix [157]. Iglesias et al. [158] prepared epoxy composites reinforced with GF thermally conditioned at 450 °C for 1 h. The effect of coating the GF with various silane coupling agents on composite tensile strength was investigated. Iglesias et al. [195] obtained no increase in composite strength when applying any of the silanes investigated, suggesting that the IFSS was not increased as a result. It was proposed that low amine group density reduced the contribution of chemical bonding to the overall adhesion; however, this was never confirmed through direct measurement or observation.

While many authors use thermally conditions GF (heat treated dry fibres) as a proxy for thermally recycled rGF (thermally reclaimed from composite materials) [157, 158]; for studies investigating IFSS changes induced by thermal recycling, it may be prudent to instead utilise rGF thermally reclaimed from actual composite materials. It is not fully understood how the direct exposure to polymer degradation influences the fibre surface chemistry, morphology, or functionality; or how this impacts the interaction with functional groups within sizings and polymers. Pender [66] measure the tensile and flexural properties of PP reinforced with fluidised bed-recycled rGF, investigating the influence of MAPP modification and fibre resizing using APS. In contrast to Nagel [157], Pender [66] found no significant improvement in strength of PP composites with the use of APS sizing alone. The combination of MAPP modification and APS sizing, however, yielded a 37% increase in PP composites using thermally recycled rGF, which was attributed to improved adhesion between MAPP and rGF. Pender [66] observed an increase in IFSS between thermally recycled rGF and epoxy with the use of APS coupling agent; however, no improvement in composite strength was realised. Kennerley et al. [71] investigated the use of methacryloxypropyltrimethoxysilane (MPS) coupling agent on tensile and flexure properties of UPR-based BMC reinforced with fluidised bed-recycled rGF. Kennerley et al. observed no discernible improvement in mechanical properties when using MPS, suggesting poor adhesion across the fibre-silane-polymer interfacial region.

Studies have observed a reduction in hydroxyl group concentration on the surface of GF after exposure to typical recycling temperatures [159, 160]. It is suggested that the dehydroxylation occurs when surface silanol groups condense to form siloxane bonds. Silane coupling agents are used in the production of GFRP bond to the glass via interaction with the surface hydroxyl groups. Low hydroxyl density on the surface of rGF may therefore lead to a weaker interfacial adhesion between the glass and polymer matrix, ultimately compromising the mechanical properties of GFRP based on thermally recycled rGF. Unfortunately, there is limited research in the area, and it remains unclear to what extent dehydroxylation occurs during thermal recycling, or whether this inhibits interaction between rGF surface and coupling agents. A promising route to market for rGF is in compounded, short fibre composites which are stiff rather than strength-limited. It is well understood that the interfacial adhesion has a larger effect on the strength of composite reinforced with shorter fibres;

therefore the fibre-polymer adhesion may play a dominant role in the strength of GFRP prepared using rGF. With no real consensus in the literature, more research is needed to better understand the impact of thermal recycling on rGF surface functionality and how it can be reliably regenerated for different polymer systems.

Several authors have reported the impact strength of GFRP reinforced with thermally recycled rGF, with data compiled in Table 3.5. Li et al. [42] observed up to 41% increase in PP unnotched impact strength when loaded with thermally recycled rGF at 30 wt.% compared to neat polymer counterpart. Impact strength was found to increase with rGF length distribution in the GFRP specimens. Similarly, Gopalraj and Karki [161] observed a 17% increase in impact strength of epoxy when loaded with thermally recycled rGF at 60 wt.% compared to neat epoxy. Impact strength was found to increase with rGF loading within the range analysed. While the addition of rGF can increase the impact strength of neat polymers, Cunliffe and Williams [4] and Kennerley et al. [71] reported reductions in impact strength when replacing virgin GF with rGF. Kennerley et al. found that impact strength of GFRP decreases as replacement rate of rGF increases; reporting a 69% reduction in impact strength when reinforced with rGF alone. From the literature it is understood that impact strength of GFRP is dependent on the fibre length, fibre content, and the IFSS [153, 162]. The reduced GFRP impact strength observed by Cunliffe and Williams [4] and Kennerley et al. [71] is therefore likely attributable to the degradation in rGF strength and/or poor fibre-polymer adhesion following thermal recycling.

3.5.3 CFRP reinforced with recycled carbon fibres

A potential route to market for thermally recycled rCF is used as reinforcement in moulding compounds. These routes avoid many of the challenges associated with handling loose and discontinuous recycled fibres by compounding them into polymers and then consolidating using compression- or injection-moulding. Turner et al. [164] prepared epoxy-based BMC and SMC reinforced with fluidised bed-recycled rCF, reporting superior tensile and flexural properties compared to virgin GF equivalent compounds. Impact strength was found to be lower than virgin GF-reinforced BMC/ SMC, which was attributed to using shorter fibre lengths and unbundled natural of thermally recycled rCF. Despite this, it was concluded that there are many applications where compounds based on thermally recycled rCF could be competitive, or superior, to virgin GF counterparts.

Several authors have reported the mechanical properties of injection-moulded composites reinforced with pyrolysis [55, 127] and fluidised bed-recycled rCF [130, 135–138]. The mechanical performance (tensile properties, flexural properties, impact strength) of injection-moulded composites fibre has been shown to generally increase with recycled fibre content [55, 127, 135–138].

Table 3.5: Properties of GFRP reinforced with thermally recycled rGF.

Matrix material	Composite type/preparation method	Recycling method	Thermal history	Virgin fibre content	Property	Recycled GFRP strength		Recycled GFRP modulus		Impact strength		Ref.
						MPa	Retained	GPa	Retained	kJ/m²	Retained	
Polyester	BMC, compression mould (150 °C, 14 MPa, 2 min)	Fluidised bed	450 °C <60 min	0%	Tensile	15	58–68%	14.1	107%	4.3	31%	[71]
				25%		17	72%	15.9	121%	10.9	77%	
				50%		25	104%	14.8	113%	10.9	78%	
				75%		21	89%	14.6	112%	12.6	90%	
				0%	Flexural	46	48%	7.6	93%			
				25%		57	60%	7.2	89%	/		
				50%		86	90%	8.0	98%			
				75%		87	91%	7.9	97%			
Polyester	BMC	Pyrolysis	450 °C 90 min	75%	Flexural	89	93%	7.3–8.2	81%	11	73%	[4]
Polyester	Hand layup, compression mould (170 °C, 4 MPa, 8 min)	Microwave pyrolysis	Non-isothermal 440 °C max	75%	Flexural	157	83%	12.4	77%			[89]
				50%		104	55%	8.6	53%	/		
				0%		61	32%	6.4	40%			
Polyester	BMC	Pyrolysis	500 °C 30 min	85%	Tensile	24	/	/	/	16.5	/	[132]
					Flexural	69	/	11.1	/			

Material	Description	Process	Temperature	Fibre	Test	Strength	%	Modulus	%			Ref
Epoxy	Infused non-woven mat (papermaking, VARTM)	Fluidised bed	500 °C	0%	Tensile	62	/	6.5	/	/	/	[66]
PP	Co-mingled non-woven mat (papermaking, compression mould, 200 °C)	Fluidised bed	500 °C	0%	Tensile	27	42%	3.5	84%	13.3	/	[66]
				50%		47	72%	4.6	110%		/	
				0%	Flexural	45	/	2.5	/		/	
PP (w/ MAPP)	Co-mingled non-woven mat (papermaking, compression mould, 200 °C)	Fluidised bed	500 °C	0%	Tensile	44	48%	4.4	94%	13.3	–	[66]
				50%		60	65%	/			/	
				0%	Flexural	59	/	2.4	/		/	
PP (w/ MAPP)	Extruded and injection-mould, 200 °C, 80 MPa	Pyrolysis	Non-isothermal, 500 °C max	0%	Tensile	33–44	/	/	/	5.5–8.2	/	[42]
					Flexural	40–53		2.2–2.6			/	
Epoxy	Hand-layup, compression-mould, UD	Cone calorimeter	550 °C	0%	Tensile	65–115	/	27–31	/	19–41	/	[161]
Epoxy	VARTM, UD	Combustion furnace	565 °C	0%	Tensile	120	9.9%	42.5	85%	/	/	[107]
Epoxy	Hand lay-up, woven	Pyrolysis	450 °C	0%	Flexural	146	/	11.7	/	/	/	[163]
Polyester						161	65%	12.3	90%	/		

Yan et al. [137] investigated the impact of fluidised bed-recycled rCF loading (5%, 10%, 15%, and 20% FWF) on PC/ABS copolymer. While tensile and flexural properties increased with rGF loading, the maximum impact strength was observed at 15 wt.%; where subsequent increase in fibre loading resulting in poorer impact properties. Similarly, Fernández et al. [127] observed improved tensile and flexural properties of injection-moulded PP when pyrolysis-recycled rCF loading was increased from 10 to 30 wt.%. It was concluded that both stiffness and strength of the composites increased linearly with the rCF fraction in a ratio of 180 and 0.9 MPa/%FVF, respectively. In agreement with Yan et al. [137], Fernández et al. [127] observed a peak in impact strength occurring at 20 wt.%, which was below the highest fibre loading evaluated. Feng et al. [135] observed an increase in tensile, flexural, and impact properties of nylon 6 compounded with fluidised bed-recycled rCF up to fibre loadings of 20 wt.%. Subsequent fibre loadings of 25 wt.% resulted in a reduction in mechanical performance (except tensile modulus). Feng et al. [135] proposed that aggregation of the recycled rGF may occur in the case of high loading and, thus, leads to regions of destroyed continuity of the matrix. This results in the decrease of crack propagation capability and severe matrix deformation. Furthermore, the agglomeration of rCF may also cause stress concentration in the polymer matrix and prevent efficient load transfer to the matrix.

Han et al. [138] reported the novel use of bio-degradable PBS in the production of CFRP using melting extrusion. An approximate linear relation between tensile, flexural, and impact properties was observed across rCF loadings of 3–20 wt.%. Han et al. [136] also used fluidised bed-recycled rCF in the production of bio-degradable-based CFRP, reporting the mechanical performance of reinforced PLA. As with other studies, the tensile, flexural, and impact properties tended to increase with rCF loading; however, impact strength was found to peak at 11 wt.%, citing aggregation of the rCF as the cause for poorer impact properties at higher fibre loadings [136].

There are a limited number of studies which produce virgin CF compounds to baseline against those reinforced with rCF counterparts. It is therefore not conclusive how the use of rCF impacts the performance of injection-moulded CFRP when substituting standard virgin CF. Many studies use commercially available rCF products which contain fibres reclaimed form feedstock of disparate origins and likely composed of various types of CF. This makes selecting an appropriate virgin CF to compare against challenging, especially given the broad range of CF grades available. Stoeffler et al. [55] compared the tensile, flexural, and impact properties of polyphenylene sulphide polymer (PPS) compounded with pyrolysis-recycled rCF and virgin CF. Since the grade of rCF was unknown, the performance of the rCF was compared to that of typical industrial grade CF (Panex 35 short CF). It was demonstrated that PPS composites reinforced with rCF (20% and 40% FWF) exhibit similar or better mechanical properties than equivalent commercial compounds produced using industrial grade short virgin CF. It was postulated by Stoeffler et al. that PPS composites made from rCF (when compared to virgin CF) might exhibit differences in (i) dispersion and/or distribution of the CF in the matrix and (ii) adhesion at the PPS/rCF interface. Analysis of SEM micrographs taken of polished longitudinal section of the prepared CFRP coupons did not show a qualitative difference between virgin and recycled CF specimens; however, orientation was not quantitatively measured. Fracture surface analysis was conducted on recycled fibre CFRP specimens. Stoeffler et al. concluded that good adhesion was generally observed at the fibre/matrix interface despite the absence of any sizing on the surface of the fibres. No comparison was made against the virgin fibre CFRP specimens; therefore, it is unknown whether the increase in CFRP performance when using rCF is attributable to greater fibre/matrix adhesion.

To date, there has been limited research comparing the effects of thermal recycling conditions (atmosphere, temperature, time) on the resulting mechanical properties of composites reinforced with thermally recycled rCF. With many studies showings the impact of thermal recycling conditions on fibre mechanical properties (Table 3.4) and adhesion with new polymers, it clear that a greater understanding of how these observations translate to composite performance is needed to find optimal recycling and reuse routes for rCF. Onwudili et al. [28] studied the impact of oxidising pyrolysis recycled rCF and its effect on mechanical properties of low-density polyethylene (LDPE) composites. Onwudili et al. [28] concluded that LDPE composite strength is reduced when using oxidised fibres, whereas tensile strength is enhanced by the oxidation stage. Onwudili et al. [28] produced rCF-reinforced LDPE composites with fibre volume fractions of 15% and measured the retained tensile, flexural, and impact properties compared to a commercial equivalent using virgin fibres. The fibres were reclaimed through pyrolysis with a nitrogen atmosphere at 500 °C. Onwudili et al. [28] also investigated the impact of secondary oxidation of the fibres on the LDPE properties (500 °C for 30 min in air). The reinforced composites were compounded using a two-roll mill, where molten LDPE polymer was blended with fibres and surface-modifying agents. The resulting product was then ground and used to manufacture composite sheets using a heated press. Tensile and flexural strength of pyrolysis-recycled rCF-LDPE composites was found to be on par with or greater than those made with virgin CF. The inclusion of the secondary oxidation stage resulted in a reduction in strength compared to rCF reclaimed using pyrolysis only as shown in Figure 3.15. This may be attributable to additional weakening of the rCF following oxidation. No micromechanical properties of rCF were measured, and the cause for the observations was not established by Onwudili et al. [28]. Flexural modulus showed similar behaviour to the strength properties; however, Onwudili et al. observed a significant increase in LDPE composite tensile modulus retention when reinforced with oxidised fibres (150%) compared to nonoxidised recycled fires (91%). No mechanistic conclusions were made by Onwudili et al. to explain these observations, and it remains unclear whether these findings are translatable to other polymer types.

Several authors have studied the use of polymer modification [55, 127, 130], and sizings/coupling agents [135, 136, 165], in the preparation of CFRP with thermally recycled rCF. It is proposed that the use of coupling agents can increase fibre-polymer

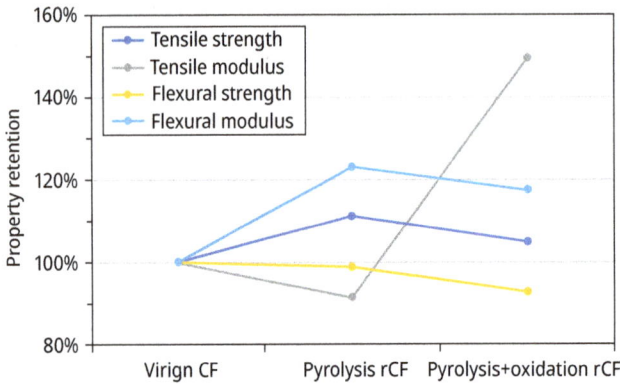

Figure 3.15: Retained mechanical properties of LDPE when reinforced with pyrolysis-recycled rCF, where virgin fibre LDPE composite properties are as follows: tensile strength – 12.6 MPa, tensile modulus – 530 MPa, flexural strength – 8.3 MPa, flexural modulus – 663 MPa (produced using data reported in [28]).

adhesion, resulting in greater performance of CFRP using rCF. MAPP has been commonly used in the production of thermoplastic-based composites reinforced with thermally recycled rCF, with studies concluding MAPP can significantly enhance resulting composite mechanical properties [28, 127, 130]. The use of silane coupling agents [136, 138], and polymer film formers [135, 137], applied directly to the surface of thermally recycled rCF have also been shown to provide desirable fibre-polymer interfacial bonding in CFRP.

Work by Shi et al. [165] investigated the effect of *N*-methyl-2-pyrrolidinone (NMP) coupling agent on the bending properties of epoxy-based composites reinforced with rCF reclaimed using steam pyrolysis. rCF were soaked in acetone for 5 days at room temperature and in NMP for 3 days at 200 °C, respectively, before finally cleaned by an ultrasonic washing machine for 1 h. Shi et al. concluded that the bending strength retention of rCF-epoxy composites could be significantly improved (from 49% to 78%) by treating rCF with NMP [165]. Shi et al. proposed that the increase in bending strength was a result of increased rCF-epoxy adhesion; however, it was not determined whether this resulted from chemical interaction with NMP coupling agent. SEM micrographs of rCF (without cleaning) showed significant surface contamination which may reduce rCF surface functionality. Cleaning rCF in acetone alone produces higher bending strength; however, this remained lower than for specimens treated with NMP. It is therefore likely that the mechanism for the observed increase in bending strength is multifaceted and a combination of both physical and chemical effects.

Feng et al. [135] investigated the effect of rCF surface treatments in the preparation of nylon 6-based thermoplastic composites reinforced with fluidised bed-recycled rCF. Feng et al. developed an effective approach to clean and modify the surface functionality of the rCF using nitric acid and epoxy macromolecular coupling agent. The rCF were first immersed in a concentrated solution of nitric acid for 2 h to clean and polarise the fibre surfaces. Fibres were immersed in a solution of bisphenol A digly-cidyl ether (DGEBA) and acetone at a concentration of 8 wt.% for 5 h and then baked under a vacuum at 90 °C for 2 h. Tensile, flexural, and impact properties were found to significantly increase when the surface treatment was used (see Table 3.6). This was attributed to improve interfacial boding between fibres and matrix, which was assessed qualitatively using fracture surface analysis. SEM micrographs did show distinct residues on the rCF surface (prior to surface treatment), which were identified as carbonaceous deposition resulting from the thermo-oxidative decomposition during the recycling process. Feng et al. proposed that these carbonaceous deposits reduce the capability of interfacial bonding with the polymeric matrix. SEM micrographs taken after soaking in nitric acid showed that this cleaning treatment alone could yield clean fibres [135]. No CFRP were prepared with rCF treated only with nitric acid; as such, it remains unclear what contributions the nitric acid cleaning alone has on the improve rCF-polymer adhesion, and whether surface modification with DGEBA is required. Yan et al. [137] also used DGEBA treatment when compounding and moulding PC/ABS composites reinforced with rCF. Yan et al. concluded that the approach is effective to improve the interfacial bonding between the thermally recycled rCF and the PC/ABS matrix.

Onwudili et al. [28] investigated the influence of different surface modifying agents on LDPE composites reinforced with pyrolysis-recycled rCF. In order to improve fibre-polymer adhesion, APS, HDPE-grafted maleic anhydride, and a polyalkenyl-polymaleic-anhydride-derived coupling agent were added into the LDPE during compounding, and their effect on mechanical properties was reported. With no coupling agents, the rCF-HDPE closely matched or outperformed the mechanical properties of the unsized virgin CF composite. The retained tensile strength and modulus was 105% and 149%, respectively, and the retained flexural strength and modulus were 93% and 117%. The polyaklenyl-polymaleic-anhydride derivative (at 2 wt.%) was found to have the maximum improvement on the tensile strength of the recycled composite, giving a mean value of 16.9 MPa. Using the best result from the additive variants, the retained properties of the recycled composite were 71% for tensile strength, 82% for tensile modulus, 99% for flexural strength, and 85% for flexural modulus [28].

Wong et al. [130] also investigated the use of different MAPP grades in injection-moulded PP composites reinforced with fluidised bed-recycled rCF. Improvement in the mechanical properties was observed with the introduction of MAPP which was attributed to the formation of better interfacial adhesion. Wong et al. determined that the compatibility was strongly dependent on the molecular weight and anhydride groups in the coupling agent. MAPP with high anhydride content was required for an efficient improvement in tensile and flexural strengths, but the maximum strength was determined by the molecular weight [130].

Several authors have reported the mechanical properties of CFRP produced using non-woven-recycled fibre intermediates with epoxy [142, 145, 166, 167], PA6/PA66 [140, 144], PET [140], and PP [142] polymer matrices. In general, composites made with non-woven-recycled fibres exhibit higher mechanical properties than those made with compounding- and injection-moulding (Table 3.6). While fibre length was not consistently measured across studies, it is probable that the average fibre length in composites of non-wovens is longer than injection-moulded counterparts, given the significant fibre length degradation caused during compounding. This may be key factor influencing the apparent higher performance of composites produced using non-wovens; however, it is difficult to make firm conclusions across studies when considering the wide range of different recycling methods, polymers, coupling agents, fibre content, and manufacturing techniques used across the literature.

A limitation of non-wovens is that they are typically produced as randomly orientated mats, therefore lacking directional anisotropy and typically have a lower maximum fibre volume fraction compared to virgin textiles. Shah and Schubel [145] investigated the mechanical properties of epoxy-based composites using randomly orientated and partially aligned fluidised bed-recycled rCF. A proprietary wet deposition process was used to prepare the non-wovens, which were infused with epoxy using VARTM. Partial alignment was found to increase VFV from 26% to 34%, resulting in tensile strength increasing from 100 to 174 MPa compared to randomly orientated counterparts. As such, Shah and Schubel [145] demonstrated that partial alignment could yield significantly improved performance of composites prepared with thermally recycled non-woven fabrics.

Several authors have investigated methods for re-aligning short thermally recycled rCF and reported the effect on CFRP properties [146, 147]. Longana et al. used thermally recycled rCF in the HiPerDiF process to produce highly aligned tapes. The fibres were reclaimed in air atmosphere at 500 °C for 5 h and then washed in an ultrasonicator for 5 min in a solution of water and 20% acetone by volume. The fibres, of lengths up to 3.5 mm, were then put through a wet alignment process with resin film applied to produce a partially impregnated tape. The tape was then laid up as 150 × 5 mm tensile test specimens and cured in an autoclave under vacuum pressure. The first relevant observation was that the mean value of fibre volume fraction of the recycled fibre HiPerDiF specimens was 28.3% compared to 37.7% for virgin. The tensile modulus and strength were normalised to account for the reduction in FVF. Without normalisation, 79% of the modulus was retained and 63% of the strength. With FVF normalisation; however, Longana et al. concluded that the recycled specimens retained 105% of the modulus and 84% of the strength. This is generally similar to the results seen on at a single fibre level in Table 3.4.

The TuFF process at University of Delaware uses a similar alignment technique to produce sheets pre-impregnated with resin film [147]. The process also disperses the short fibres, 3 mm in length, in a mixing tank followed by sonication before going through multiple alignment steps. Heider et al. manufactured coupons by stacking 24 plies of the 8 gsm output from the TuFF-reformatting process. The resulting fibre volume fraction achieved was 48%, significantly higher than was reported by Longana et al. and more comparable to traditional virgin CF composites. The source of the rCF in this case was Toray T800H prepreg material. The rCF were initially reclaimed by solvolysis, but Heider et al. observed via SEM significant residue remaining on the fibres and therefore put the fibres through a purely thermal pyrolysis process. They used an inert nitrogen atmosphere and held a temperature of 500 °C for a dwell time of 60 min. The resulting mechanical data for the TuFF-recycled coupons were compared to the Toray T800H datasheet with 60% fibre volume fraction. Compared to the datasheet without normalisation of fibre volume fraction, the TuFF-recycled composite retained 43% of the tensile strength and 78% of the modulus. When normalising the T800H datasheet properties to 48% fibre volume fraction, the retention for strength and modulus increased to 54% and 98%, respectively.

Many authors have also produced hybrid composite materials reinforced with pyrolysis-recycled rCF and flax [166, 168, 169] and hemp [170, 171] fibres. Hybrid composites have been prepared in several ways including resin infusion [166], compounding and injection-moulding [171], rotational-moulding [170], wet laid [169], and as realigned tapes [168]. Shah et al. reported no significant reduction in tensile strength of rCF-reinforced PP composites, when replacing 50 wt.% of rCF with hemp. Moreover, the tensile strength of the hybrid composite was found to be around 35% stronger than hemp fibre-only specimens. Tensile modulus of the hybrid composites was observed to be lower than both flax fibre and rCF-only composites. Oliveira et al. [170] reported the mechanical properties of hybrid PET composites reinforced with hemp and thermally recycled rCF. It was discovered that the use of rCF significantly increased both tensile strength and modulus compared to hemp fibre-only composites. Wilson et al. [166] produced flax/rCF-reinforced epoxy hybrid composites. Woven flax fabrics were stacked with thermally recycled and carded rCF non-woven mats. It was found that a higher fibre volume fraction could be achieved with the integration of the flax-woven fabric. The hybrid composite was found to have significantly improved the tensile properties compared to flax fibre-only and were comparable to those reinforced with rCF alone [166]. Similarly, Tse et al. [171] prepared flax/rCF hybrids (using wet laid) with biodegradable PLA thermoplastic. Tse et al. [171] observed flexural strength to increase by 60% and 81% when replacing 25 and 50 wt.% of flax fibre with pyrolysis-recycled rCF. Similar trends were also observed for flexural modulus.

Longana et al. [168] demonstrated the production of aligned flax/rCF tapes. The tensile strength and modulus of composites prepared with the hybrid tapes were found to fall linearly with flax fibre content, concluding that hybrid materials can be a viable solution in applications, where a reduction of primary mechanical properties is an acceptable trade-off for an increase of functional properties (such as dampening) and cost reduction. In all studies, it was therefore found that the inclusion of pyrolysis-recycled rCF can yield significant performance gains compared to natural fibre-only counterparts. While still relatively new, these hybrids have the potential to accelerate the use of natural fibres into more advanced and demanding applications by overcoming the limited mechanical performance of traditional natural fibre composites. Hybridisation will likely add additional complexity and challenges for these materials at end of life which has yet to be studied in the literature.

Table 3.6: Properties of CFRP-reinforced with thermally recycled rCF, [1]notched Izod impact test – 1/8 in thickness, [2]depending on coupling agent used and loading between 2 and 8 wt.% [130], [3]notched Izod impact test – 3.17 mm thickness, and [4]range dependent on orientation of coupon extracted from CFRP (0°, 90°).

Matrix material	Preparation method	Fibre content	Recycling method	Recycling thermal history	Property	Recycled strength		Recycled CFRP modulus		Impact strength		Ref.
						MPa	Retained	GPa	Retained	Unit	Retained	
Epoxy	BMC	10% FVF	Fluidised bed	Unspecified	Tensile	71		20		8 kJ/m^2		[164]
					Flexural	98		18				
					Compression	257	/	/	/	/		
	SMC	23% FVF			Tensile	282		44				
					Flexural	520		32		32 kJ/m^2	/	
					Compression	523		/				
Nylon 6	Compound + injection-moulding	5–25% FWF	Fluidised bed	Unspecified	Tensile	73–120		8–13		49–54 J/m[1]		[135]
					Flexural	130–185	/	8.4–11	/		/	
	rCF surface unmodified											
	Injection-moulded				Tensile	105–170		11–20		62–105 J/m[1]		
					Flexural	165–240		11–14				
	rCF surface-modified w/ DGEBA											

Material	Process	FWF	Method	Temp	Property							Ref
PP/MAPP	Compound + injection-moulding	10–30% FWF	Pyrolysis	Unspecified	Tensile	45–63	/	3.5–7.0	/	33–38 kJ/m²	/	[127]
					Flexural	68–87		3.2–8.6				
PP	Compound + injection-moulding	30% FWF	Fluidised bed	550 °C	Tensile	50	/	14.5	/	13 kJ/m²	/	[130]
					Flexural	90		14				
PP/MAPP					Tensile	80–125		15–18		9–33 kJm²		
					Flexural	130–220		13.5–16.5				
Polyphenylene sulphide	Compound + injection-moulding	20–40% FWF	Pyrolysis	400 °C	Tensile	146–191	110–136%	17–30	99–118%	38–45 J/m³	115–131%	[55]
					Flexural	233–325	125–140%	17–29	118–134%			
Polylactic acid	Compound + injection-moulding	3–15% FWF	Fluidised bed	Unspecified	Tensile	76–110	/	5.4–11.5	/	27–42 J/m¹	/	[136]
					Flexural	112–144		3.2–8.1				
PC/ABS (80/20) copolymer	Compound + injection-moulding	5–20% FWF	Fluidised bed	Unspecified	Tensile	91–136	/	8.7–19	/	179–249 J/m¹	/	[137]
					Flexural	140–219		7.1–15				
PC/ABS (70/30) copolymer					Tensile	88–130	/	7.0–17	/	190–252 J/m¹	/	
					Flexural	132–196		6.3–14				
PC/ABS (60/40) copolymer					Tensile	88–131		6.0–17		198–263 J/m¹		
					Flexural	132–190		6.0–12				
Poly(butylene succinate)	Compound + injection-moulding	3–20% FWF	Fluidised bed	Unspecified	Tensile	36–56	/	1.3–6.2	/	31–58 J/m¹	/	[138]
					Flexural	34–75		0.76–4.8				

(continued)

Table **3.6** (continued)

Matrix material	Preparation method	Fibre content	Recycling method	Recycling thermal history	Property	Recycled strength		Recycled CFRP modulus		Impact strength		Ref.
						MPa	Retained	GPa	Retained	Unit	Retained	
PA6	Carding + compression-moulding	33% FVF	Pyrolysis	Unspecified	Flexural	336–358	/	20–21	/		/	[140]
PET		30% FVF			Flexural	355	/	26	/			
Epoxy	Wet laid + RTM	25% FVF	Pyrolysis	Unspecified	Tensile	186–229[4]	/	19–23[4]	/		/	[142]
PP	Carding + compression-moulding	20% FVF			Tensile	64–77[4]	/	9.4–11.5[4]	/		/	
Epoxy	Carding + VARTM	17% FVF	Pyrolysis	Unspecified	Tensile	145	/	12	/		/	[166]
PA6/PA66 copolymer	Wet laid + compression-moulding	25% FVF	Fluidised bed	Unspecified	Tensile	136–148	/	14–17	/		/	[144]
					Flexural	278–287	/	15–16	/		/	
Epoxy	Wet laid + compression-moulding	30% VFV	Pyrolysis	Unspecified	Tensile	117–195	/	16–28	/		/	[167]
Epoxy	Wet laid (Random mats) + VARTM	25–26% FVF	Fluidised bed	Unspecified	Tensile	84–100	/	12–13	/		/	[145]
	Wet laid (Partially aligned mats) + VARTM	27–34% FVF			Tensile	133–174	/	19–32	/		/	

LDPE	Two-roll mill + Hot press	15% FVF	Pyrolysis	N_2 500 °C 0 min dwell + oxidation	Tensile	13.2	105%	0.79	149%	[28]
					Flexural	7.7	93%	0.78	117%	
LDPE w/ coupling agent				500 °C 30 min	Tensile	16.9	71%	0.97	82%	/
					Flexural	16.1	99%	1.21	85%	
Epoxy	VARTM of woven fabric	46% FVF	Steam pyrolysis	340 °C 30 min	Compression	145	65.2%			[172]
Epoxy w/ NMP coupling agent					Compression	194	87.5%	/	/	/
Epoxy	Resin film infusion of woven fabric	50–5% FVF	Pyrolysis	Unspecified	Tensile	271–634	/	56–66	/	[148]
Epoxy film	HiPerDiF alignment technology	28% FVF	Pyrolysis	500 °C in air	Tensile	461	63%	52	79%	[146]
Epoxy film	TuFF alignment technology	48% FVF	Pyrolysis	500 °C in N_2 60 min	Tensile	1269	43%	132	78%	[147]

3.6 Progress toward commercial composite thermal recycling

3.6.1 GFRP recycling

Much of the commercial activities in GFRP thermal recycling are focused on solutions for scrap composite wind turbine blades; with cement kiln co-processing and pyrolysis being the dominant approaches being developed at scale. Pyrolysis-based thermal recycling approach for wind turbine blades has been adopted by several organisations including Carbon Rivers (USA), Gjenkraft (Norway), MAKEEN Energy (Denmark), Bcircular (Spain), and Refiber APS (Denmark).

Co-processing of GFRP in cement kilns is well-established and demonstrated to be a commercially viable option to re-direct scrap wind turbines away from landfill and waste to energy facilities. Neocomp (Germany) is a joint venture between Neowa (Germany) and Nehlsen International (Germany), established in 2015 [173]. Neocomp is also part of the European brand Fibreglass Recycling Europe which provides collection, transport, and residue-free recycling of GFRP waste as feedstock to the cement industry. Customers of Fibreglass Recycling Europe can use the Fibreglass Recycling Europe label to mark environmentally friendly and sustainable recycling of their products. The brand is supported by the interest group representing the plastics industry in Germany, AVK (Germany), Fiberline Composites (Denmark), and the European Composites Industry Association, EuCIA (Belgium). In addition to wind blade waste, Fibreglass Recycling Europe purports to accept other forms of GFRP production waste as well as GFRP dust collected during production [173].

The Danish multinational FLSmidth (Copenhagen, Denmark) [174] is part of the consortium behind the DecomBlades project with Ørsted, LM Wind Power, Vestas Wind Systems A/S, Siemens Gamesa Renewable Energy, MAKEEN Power, HJHansen Recycling, and Energy Cluster Denmark (ECD) [175]. FLSmidth is investigating the possibilities of using shredded blade material and pyrolysis products in the cement production process. As a knowledge and technology provider to the cement industry, FLSmidth's main objective within the DecomBlades project is to evaluate possible solutions for incorporating blade materials in cement production on a global scale. The proprietary pyrolysis process of MAKEEN Energy, Plastcon®, is currently in use in the DecomBlades project [176]. MAKEEN Energy have taken prior knowledge of the established Plastcon® pyrolysis technology to design a pilot plant that, it claims, will soon be able to operate in a continuous process for GFRP recycling with the end goal to "design a full-scale pyrolysis plant for composite materials which interested parties can order" [177].

Swiss waste management organisation Geocycle is a partnership between Zajons Zerkleinerungs GmbH (Germany-based recycler) and Holcim-Lafarge which uses GFRP waste from wind turbine blade to produce clinker in cement kilns. Geocycle launched operations in UK in 2017; however, they are yet to establish a similar supply change for waste GFRP in the UK.

In late 2020, GE Renewable Energy signed an agreement with Veolia North America, VNA (USA) to recycle its onshore composite wind blades in the United States [178, 179]. This recycling contract is the first of its kind in the United States, which aims to use the cement kiln co-processing solution that has already proven its effectiveness in Europe. As part of the agreement, GFRP blades will be shredded at a VNA processing facility in Missouri and then used as a replacement for coal, sand, and clay at cement manufacturing facilities in the United States. On average, nearly 90 wt.% of the composite blade material will be reused as a repurposed engineered material for cement production, with more than 65 wt.% replacing raw materials that would otherwise be added to the kiln to create the cement.

Researchers at University of Tennessee and partners Carbon Rivers purports to have developed a novel, multi-stage pyrolysis technology capable of separating polymers and other organic materials from GFRP, apparently allowing the GF to be recovered mechanically intact and then reused for manufacturing [180]. Pyrolysis gases are captured as energy to power the recovery process, reportedly minimising input energy requirements. The company has expanded its capacity at its Knoxville facility, including a 2.5-acre laydown yard for accepting incoming wind blade and other composite waste material streams. A new 26,000-ft^2 facility retains a 30 ton/month pilot-recycling line and 40 ton/month thermoplastic pellet production capacity. As of 2021, Carbon Rivers operates a 1 ton/day process, but by 2023 aims to achieve 200 ton/day in multiple facilities located in the USA, Europe, and Asia.

Bcircular is a spin-off of the Spanish National Research Council (CSIC) and worked on the reclamation of the GF from the damaged wind blades of the Bon vent Vilalba wind farm (EDP renewables) [181]. The pyrolysis process has been validated on the pilot scale within the R3Fiber project, backed by Horizon 2020 funding. It is claimed that "A highly energy-efficient, closed process allows obtaining high quality rGF and rCF, together with sub-products that can be re-used as fuel within the process itself, or to produce heat and energy". It is not clear at what scale Bcircular technology has been demonstrated or if the organisation actively recycles FRP waste commercially.

ReFiber ApS was funded by Nordisk Aeroform ApS in 2002 supported by Teknologisk Innovation of Copenhagen [182]. GFRP wind blade sections have been recycled using a 6-m pilot-line pyrolysis rotary kiln. After the size reduction, the material is fed into an oxygen-free rotating furnace at a temperature of 500 °C. Gas produced during the pyrolysis is used for electricity production as well as heating the furnace, and the fibres are cleaned in a secondary combustion process. Due to strength loss, the rGF can only be used in non-structural applications such as thermal-resistant materials (e.g. wool insulation).

Gjenkraft AS recycling technology is based on a variation of the pyrolysis process which is reported to enable the recovery of rGF and rCF from FRP waste as well as polymer fraction in the form of synthetic oils and fuels. Gjenkraft's reported target markets include composite from turbines blades, marine, automotive, aviation, and construction. 2022 saw Gjenkraft partner with Swedish multinational power company Vattenfall as part of the operators target to recycle at least 50% of all decommissioned turbine blades by 2025 [183]. As part of this initiative, Gjenkraft aims to demonstrate recycling of blades from their decommissioned Irene Vorrnik onshore wind farm and produce recycled fibres, synthetic oils, and gas; fractions of which can be reused in production of sporting equipment (such as skis and snowboards) and insulation materials. In 2022 wind turbine OEM Vestas introduced a ski, which is made from rCF recycled from Vestas blades, made by Gjenkraft, EVI Ski, Vest Resirkcenter, and Fred Olsen Renewables in Norway [184]. It remains unclear what capacity Gjenkraft AS is currently operating, and how many of the 84 turbine blades being decommissioned in Vattenfall's Irene Vorrnik site will recycle using Gjenkraft pyrolysis technology. The performance and potential routes to market for rGF and synthetic organics recovered from said blades has also not yet been reported.

Cubis Systems (UK) is co-financing the up-scaling and commercialisation of thermal recycling of GFRP using the fluidised bed·process [185]. Work began in 2022 with a £2 million project to develop the UK's first wind turbine blade-recycling pilot plant. Led by Scottish researchers, the pilot aims to demonstrate a commercially viable solution for GFRP waste. Project partners include Cubis Systems, trade body Composites UK, University of Strathclyde's Advanced Composites Group and Lightweight Manufacturing Centre, University of Nottingham, global waste management firm SUEZ, and composite distributor GRP Solutions.

Over several years, CHZ Technologies have been in collaboration with the American Composites Manufacturing Association and the Institute for Advanced Composites Manufacturing Innovation to demonstrate at-scale thermal recycling of composite waste using their Thermolyzer™ technology [186]. The Thermolyzer uses continuous, oxygen-free, pyrolysis to break down the composite materials into rGF and/or rCF as well as a clean gas that can be used for powering the process. The Thermolyzer system operates continuously using the recoverable embodied energy in the waste materials; therefore, once started, no additional energy is reported or needed to operate the unit. The collaboration has seen the Thermolyzer™ process composite materials from four different sources: a tractor panel containing GFRP and CFRP, SMC automotive panel, a CFRP wind blade spar cap, and wind blades that contained GFRP, balsa wood, and other materials. Subsequent development has focussed on the wind sector, having demonstrated batch processing of up to 500 lb of composite wind blade waste using the Thermolyzer™ technology [187].

3.6.2 CFRP recycling

Carbon Conversions Inc. (USA) employs pyrolysis technology in the recycling of CF that can process "the entire composites waste stream" including continuous tow, dry fibre intermediates (from fabrics, trimming, braiding), uncured prepreg and pultrusion waste, and end-of-life parts from commercial aerospace, recreational, and industrial sources [186]. Carbon Conversions boasts an annual capacity of 1,800 ton of CF waste and have a range of commercially available rCF products including chopped fibres, for use in thermoplastic-compounding and injection-moulding processes, BMC and SMC, as well as non-woven mats for use in processes that include closed/open-mould infusion, prepreg, and compression-moulding. The company has successfully demonstrated use of their recycling carbon fibre products in multiple sectors including a floorboard part for an automotive OEM and tooling in marine and aerospace. In 2016, global leader in advanced composites technology, Hexcel Corp., announced that it has made a $5 million investment in Carbon Conversions [188].

Mitsubishi Chemical Advanced Materials GmbH (MCAM) is Europe's first, and most established, commercially operating thermal recycler for CF waste. Formally owned by Germany-based Karl Meyer AG, the sale of its carboNXT GmbH and CFK Valley Stade Recycling GmbH & Co. CF recycling subsidiaries to MCAM was announced in 2020 [189]. MCAM utilise pyrolysis technology to recycle materials generated during the moulding of intermediate materials such as CF prepreg, mainly from customers in mobility-related industries as well as CFRP waste streams from automotive, aerospace, and end-of-life sporting goods [186]. MCAM boasts an annual waste capacity of 1,000 ton and produces a range of rCF products distributed through its carboNXT brand. These include chopped and milled products as well as wet-laid veils and air-laid non-woven mats. It is reported that milled rCF have found route to market as filler in polyurethane rear and front bumpers on the Mercedes AMG GTC roadster [186].

Karborek RCF Srl is a composite recycler operating in Puglia, Italy, utilising pyrolysis technology to recycle CFRP. In collaboration with ENEA Grupo, Karborek RCF's pyrolysis technology was patented in 2002. A target capacity of 1,500 ton of scrap materials per year has been reported; however, it remains unclear what capacity the facility has achieved. A range of propriety rCF products, with the trade name "KARBO", have been proposed including chopped, milled and non-woven products [186].

A novel steam pyrolysis-recycling process has been developed by B&M Longworth (UK), called "DEECOM", which has been demonstrated to reclaim rCF with near virgin performance. The process uses high-temperature steam and a pressurised reactor to decompose commonly used polymer matrix materials and can reclaim rCF with 80–100% cleanliness. While the superheated steam is still present, the vessel is pressurised by at least 0.5 bar above atmospheric pressure and goes through several cycles of compression and decompression; the frequency and intensity of the cycles depend on the waste material. Through a new partnership with global equipment supplier to the composites industry, Cygnet Texkimp (UK), Longworth plans to begin selling DEECOM units in 2022 [190].

3.6.3 Recycled fibre products

There are three distinct reuse cases for thermally recycled rCF that are either done commercially or are under development, which are as reinforcement in (1) compounding materials, (2) non-woven fabrics, or (3) aligned fabrics/tapes [186, 191–193]. Both compound materials (BMC/SMC/injection-moulding) and non-wovens are commercially produced using thermally recycled rCF. Carbon Conversions Inc. and MCAM (through CarboNXT) produce and distribute moulding compound and non-woven fabric using rCF recycled using their pyrolysis lines. A range of these products are described in Table 3.7.

Extracting the most value of the rCF is economically advantageous, which can be accomplished by ensuring that fibre can be utilised in high value end products. From the commercial processes available, reformatting rCF into a non-woven is likely the superior option (compared to moulding compounds), with carding the most used route for producing rCF non-woven fabrics (Gen2Carbon, CarboNXT, Carbon Conversions). An example of commercially available rCF non-woven product is in veils (also known as carbon tissue) which are manufactured from ultra-light non-woven chopped-strand fibre which are randomly held together with a weak binder. This material is typically used to provide a smooth surface on complex composite parts or to achieve a quasi-isotropic surface conductivity [6]. These materials typically produced using wet-laid. Thicker non-woven mats or felts can be produced with either recycled fibre alone and infused with resin or co-mingled with thermoplastics and compression-moulded.

Table 3.7: Commercially available thermally recycled rCF products, typical characterises, and applications [194].

Product	Characteristics	Use/applications
Milled rCF	Fibre length 80–500 µm Fibre tensile strength > 3.5 GPa Fibre tensile modulus > 230 GPa	Filler that enhances both the mechanical and electrical properties. Applications such as plastic housings, floor coverings, and in the medical sector. Free-flowing versions also available produced by fibre compaction into balls/tubes.
Chopped rCF	3–100 mm Fibre tensile strength > 3.5 GPa Fibre tensile modulus > 230 GPa	Used as feedstock in production of BMC/SMC or rCF fleeces (veil/non-woven). They are also used in injection-moulding. Applications that span automotive, electronics, and sporting goods.

Table 3.7 (continued)

Product	Characteristics	Use/applications
rCF thermoplastic compound	Composite tensile strength: 110 MPa Composite tensile modulus: 14.5 GPa	Compounds with rCF for use in injection-moulding processes. Available with various polymers such as PP, PC, and polyamide.
rCF SMC	Epoxy-based composite tensile strength: 233 MPa Composite tensile modulus: 32.9 GPa	Applications where high mechanical strength and rigidity are required such as interior of automobiles and airplanes.
rCF BMC	Vinylester-based composite tensile strength: 208 MPa Composite tensile modulus: 19.3 GPa	Applications can be found in the construction and electrical industries as well as in vehicle construction. The material can be processed by pressing or injection-moulding.
rCF veil/paper	0–100 gsm	Typically wet-laid non-wovens, suitable for surface lamination or the reinforcement of sandwich components.
rCF non-woven mat/fleece	Up to 500 gsm	Typically carded and needle-bonded/sewn fabric, either rCF-only or blended with thermoplastic. Can be used in closed/open-mould infusion, prepreg, and compression-moulding and have application in automotive, wind energy, marine, and tooling.

3.7 Conclusions

Sustainability at end of life continues to be a growing concern for the composite industry and the sectors that use them. This has catalysed a growth toward commercialisation of thermal recycling technologies for both CFRP and GFRP over the last decade. Thermal-recycling rCF is already commercially active in both Europe and the USA; however, it appears that most operations accept feedstock in the form of production waste opposed to end-of-life CFRP. Given the prospect of a growing, and relative consistent feedstock, the scale-up and commercialisation of thermal recycling of GFRP waste has nucleated in the wind energy sector. Several high-value projects are underway to demonstrate at scale-up of thermal recycling of end-of-life wind blades.

Thermal recycling of fibre-reinforced polymer composites involves volatilising the polymer matrix at elevated temperatures of 400–700 °C. To date, research has emphasised the reclamation of the fibre, fillers, and other inserts at the expense of the matrix in its raw material form. Key parameters that are considered during thermal recycling are the processing temperature and residence time, feedstock size, atmosphere in the recycling process, accepted level of contamination, desired performance of recyclate, and desired rate of waste throughput. A multitude of thermal recycling methods have been investigated for both carbon and glass fibre polymer composite materials, which can be broadly categorised between those which combust and those which pyrolyse the polymeric matrix. Combustion processes involve heating the composite in the presence of oxygen, allowing for rapid thermal decomposition of the polymer which can facilitate low residence times and high rate of waste throughputs and produce clean fibres with low surface contamination.

Pyrolysis recycling uses indirect heat, in an oxygen-starved environment, to thermally crack the polymer fraction in waste composites. Limiting oxidation allows for the potential recovery of not only reinforcement fibres but also the polymer fraction in the form of hydrocarbon gases, liquids (oils), and solids (wax, char). Carbonised contamination on the surface of pyrolysed fibres is commonly observed; therefore, secondary cleaning processes are often used (such as oxidation, chemical, and ultrasonic cleaning) to facilitate the recovery of clean fibres. Typical atmospheric conditions used in pyrolysis include carbon dioxide, recirculated pyrolysis product gases, superheated steam, and vacuum; however, nitrogen is most commonly used in an academic setting.

Several studies have characterised the polymer products obtained from composite pyrolysis, identifying styrene as a potentially usable product from UPR resin pyrolysis. Even relatively small dosing of contaminants in the feedstock (such as PVC and halogen flame retardants) can produce undesirable chemical products which can be detrimental to the process itself as well as contaminate and deteriorate the quality of oil products. Demonstrated use of pyrolysis products of widely used thermoset resins used in FRP, such as epoxy resins, UPR, and vinyl ester resins, has yet to be reported. More work is therefore needed before determining whether polymer products from pyrolysis have a feasible route to market as anything more than fuels.

Thermally recycled fibres have historically proven challenging to integrate into composite manufacturing due to the low-density, randomly oriented, discontinuous fibre format that results from typical thermal recycling. Three distinct classes of intermediate materials have been produced using thermally recycled fibres: moulding compounds, non-woven mats, and aligned tapes/prepregs.

It is widely reported that the strength of composites produced with thermally recycled rGF or rCF is significantly lower than virgin equivalents. The drop in strength is typically attributed to one or more of the following: strength loss of recycled fibres, reduced adhesion between fibre and polymer, and/or reduced fibre content of composites made with recycled fibre.

Although rGF strength has been shown to significantly drop during thermal recycling, the tensile modulus of rGF does not degrade following exposure to typical thermal recycling temperatures. A promising route to market these recyclates is in compounded, short fibre composites which are stiff rather than strength-limited. It is well understood that the interfacial adhesion has a larger effect on the strength of composite reinforced with shorter fibres; therefore, the fibre-polymer adhesion may play a dominant role in the strength of GFRP prepared using rGF. It has been shown that fibre sizings degrade when exposed to typical thermal recycling conditions and that this can result in reduce adhesion between rGF and thermoplastics or thermosets. In some cases, the reapplication of sizing and/or polymer modification has been shown to boost adhesion with rGF and should be chased as a method for improving the strength of second-life GFRP.

The strength retention of thermally recycled rCF is general found to decrease with increasing recycling temperature, exposure time, and oxygen content present in the process. There remains no clear consensus in the literature as to whether the impact of thermal recycling is positive, negative, or neutral with respect to the adhesion between rCF and polymer systems. It is difficult to isolate the cause for the reduced strength of composites reinforced with thermally recycled rCF and is likely a combination of both fibre properties and poorer fibre-polymer adhesion. Regardless of the mechanism, the lower composite strength makes it challenging to compete with virgin materials in applications which have high load carrying capacity. Moreover, the loss in fibre alignment and length continuity means that industry standard intermediate formats (such as prepregs, NCF, and woven fabrics) are not currently available with thermally recycled rCF. In their present format, rCF are therefore not a drop in solution in the production of typical CFRP products. Alternative applications which have performance and production requirements suited to thermally recycled rCF must therefore be identified as a route to market for these unique materials.

References

[1]   L. Giorgini *et al.*, "Pyrolysis of fiberglass/polyester composites: Recovery and characterization of obtained products," *FME Trans.*, vol. 44, no. 4, pp. 405–414, 2016, doi: 10.5937/fmet1604405G.

[2]   Y. Ren *et al.*, "Evaluation of Mechanical Properties and Pyrolysis Products of Carbon Fibers Recycled by Microwave Pyrolysis," *ACS Omega*, vol. 7, no. 16, pp. 13529–13537, 2022, doi: 10.1021/acsomega.1c06652.

[3]   A. Torres *et al.*, "Recycling by pyrolysis of thermoset composites: characteristics of the liquid and gaseous fuels obtained," *Fuel*, vol. 79, no. 8, pp. 897–902, Jun. 2000, doi: 10.1016/S0016-2361(99)00220-3.

[4]   A. M. Cunliffe and P. T. Williams, "Characterisation of products from the recycling of glass fibre reinforced polyester waste by pyrolysis," *Fuel*, vol. 82, pp. 2223–2230, Dec. 2003, doi: 10.1016/S0016-2361(03)00129-7.

[5]   J. Bhadra, N. Al-Thani, and A. Abdulkareem, *Recycling of polymer-polymer composites.* Sawston, Cambridge, England: Woodhead Publishing, Elsevier Ltd., 2017.

[6]   A. J. Nagle, E. L. Delaney, L. C. Bank, and P. G. Leahy, "A Comparative Life Cycle Assessment between landfilling and Co-Processing of waste from decommissioned Irish wind turbine blades," *J. Clean. Prod.*, vol. 277, p. 123321, 2020, doi: 10.1016/j.jclepro.2020.123321.

[7]   X. Xue, S. Y. Liu, Z. Y. Zhang, Q. Z. Wang, and C. Z. Xiao, "A technology review of recycling methods for fiber-reinforced thermosets," *J. Reinf. Plast. Compos.*, vol. 41, no. 11–12, pp. 459–480, 2022, doi: 10.1177/07316844211055208.

[8] S. J. Pickering, "Recycling technologies for thermoset composite materials-current status," *Compos. Part A Appl. Sci. Manuf.*, vol. 37, no. 8, pp. 1206–1215, Aug. 2006, doi: 10.1016/j. compositesa.2005.05.030.

[9] K. Pender and L. Yang, "Regenerating performance of glass fibre recycled from wind turbine blade," *Compos. Part B Eng.*, vol. 198, p. 108230, 2020, doi: 10.1016/j.compositesb.2020.108230.

[10] S. J. Pickering, R. M. Kelly, J. R. Kennerley, C. D. Rudd, and N. J. Fenwick, "A fluidised-bed process for the recovery of glass fibres from scrap thermoset composites," *Compos. Sci. Technol.*, vol. 60, no. 4, pp. 509–523, 2000, doi: 10.1016/S0266-3538(99)00154-2.

[11] K. Pender and L. Yang, "Investigation of catalyzed thermal recycling for glass fiber-reinforced epoxy using fluidized bed process," *Polym. Compos.*, vol. 40, no. 9, pp. 3510–3519, 2019, doi: 10.1002/pc.25213.

[12] S. Francis, "Decommissioned wind turbine blades used for cement co-processing," *Composite World*, 2019. https://www.compositesworld.com/articles/recycled-composites-from-wind-turbine-blades-used-for-cement-co-processing#:~:text=7%2F18%2F2019-,Decommissioned wind turbine blades used for cement co-processing, raw material and saving energy. (accessed Apr. 01, 2024).

[13] Y. Zhang, Y. Cui, P. Chen, and S. Liu, *Gasification Technologies and Their Energy Potentials*. Amsterdam, Netherlands: Elsevier, Elsevier B.V., 2019.

[14] P. Lettieri and S. M. Al-salem, *Thermochemical Treatment of Plastic Solid Waste*. Cambridge, MA, USA: Academic Press, Elsevier Inc., 2011.

[15] P. Dwivedi, P. K. Mishra, M. K. Mondal, and N. Srivastava, "Non-biodegradable polymeric waste pyrolysis for energy recovery," *Heliyon*, vol. 5, no. 8, p. e02198, 2019, doi: 10.1016/j.heliyon.2019. e02198.

[16] N. Kiran, E. Ekinci, and C. E. Snape, "Recycling of plastic wastes via pyrolysis," *Fuel Energy Abstr.*, vol. 41, no. 6, pp. 417–418, 2000, doi: 10.1016/s0140-6701(00)94792-1.

[17] I. M. Maafa, "Pyrolysis of Polystyrene Waste : A Review," *Polymers*. vol. 13, no. 2, pp. 225, https://doi. org/10.3390/polym13020225. 2021.

[18] C. Abdy, Y. Zhang, J. Wang, Y. Yang, I. Artamendi, and B. Allen, "Resources, Conservation & Recycling Pyrolysis of polyolefin plastic waste and potential applications in asphalt road construction : A technical review," *Resour. Conserv. Recycl.*, vol. 180, no. September 2021, p. 106213, 2022, doi: 10.1016/j.resconrec.2022.106213.

[19] W. Kaminsky and J. Franck, "Monomer recovery by pyrolysis of poly (methyl methacrylate) (PMMA)," *J. Anal. Appl. Pyrolysis.*, vol. 19, no. 1991, pp. 311–318, 2000.

[20] R. Ahead, "Current Technologies in Depolymerization Process and the Road Ahead," *Polymers*, vol. 13, no. 3, pp. 1–17, 2021.

[21] E. Pakdel, S. Kashi, R. Varley, and X. Wang, "Recent progress in recycling carbon fibre reinforced composites and dry carbon fibre wastes," *Resour. Conserv. Recycl.*, vol. 166, p. 105340, 2021, doi: 10.1016/j.resconrec.2020.105340.

[22] Y. Yang, R. Boom, B. Irion, D. J. Van heerden, P. Kuiper, and H. De wit, "Recycling of composite materials," *Chem. Eng. Process. Process Intensif.*, vol. 51, pp. 53–68, 2012, doi: 10.1016/j.cep. 2011.09.007.

[23] L. Giorgini, T. Benelli, G. Brancolini, and L. Mazzocchetti, "Recycling of carbon fiber reinforced composite waste to close their life cycle in a cradle-to-cradle approach," *Curr. Opin. Green Sustain. Chem.*, vol. 26, p. 100368, 2020, doi: 10.1016/j.cogsc.2020.100368.

[24] S. Pimenta and S. T. Pinho, "Recycling carbon fibre reinforced polymers for structural applications: Technology review and market outlook," *Waste Manag.*, vol. 31, no. 2, pp. 378–392, Feb. 2011, doi: 10.1016/j.wasman.2010.09.019.

[25] L. Mazzocchetti, T. Benelli, E. D'Angelo, C. Leonardi, G. Zattini, and L. Giorgini, "Validation of carbon fibers recycling by pyro-gasification: The influence of oxidation conditions to obtain clean fibers and promote fiber/matrix adhesion in epoxy composites," *Compos. Part A Appl. Sci. Manuf.*, vol. 112, no. October 2017, pp. 504–514, 2018, doi: 10.1016/j.compositesa.2018.07.007.

[26]   L. O. Meyer, K. Schulte, and E. Grove-Nielsen, "CFRP-Recycling Following a Pyrolysis Route: Process Optimization and Potentials," *J. Compos. Mater.*, vol. 43, no. 9, pp. 1121–1132, 2009, doi: 10.1177/0021998308097737.

[27]   F. A. López *et al.*, "Recovery of carbon fibres by the thermolysis and gasification of waste prepreg," *J. Anal. Appl. Pyrolysis*, vol. 104, pp. 675–683, 2013, doi: http://dx.doi.org/10.1016/j.jaap.2013.04.012.

[28]   J. A. Onwudili, N. Miskolczi, T. Nagy, and G. Lipóczi, "Recovery of glass fibre and carbon fibres from reinforced thermosets by batch pyrolysis and investigation of fibre re-using as reinforcement in LDPE matrix," *Compos. Part B Eng.*, vol. 91, pp. 154–161, 2016, doi: 10.1016/j.compositesb.2016.01.055.

[29]   J. Yang, J. Liu, W. Liu, J. Wang, and T. Tang, "Recycling of carbon fibre reinforced epoxy resin composites under various oxygen concentrations in nitrogen–oxygen atmosphere," *J. Anal. Appl. Pyrolysis*, vol. 112, pp. 253–261, 2015, doi: http://dx.doi.org/10.1016/j.jaap.2015.01.017.

[30]   J. A. Onwudili, N. Insura, and P. T. Williams, "Autoclave pyrolysis of carbon reinforced composite plastic waste for carbon fibre and chemicals recovery," *J. Energy Inst.*, vol. 86, no. 4, pp. 227–232. 2013, https://doi.org/10.1179/1743967113Z.00000000066.

[31]   K. W. Kim, H. M. Lee, J. H. An, D. C. Chung, K. H. An, and B. J. Kim, "Recycling and characterization of carbon fibers from carbon fiber reinforced epoxy matrix composites by a novel super-heated-steam method," *J. Environ. Manage.*, vol. 203, pp. 872–879, 2017, doi: 10.1016/j.jenvman.2017.05.015.

[32]   J. Shi, L. Bao, R. Kobayashi, J. Kato, and K. Kemmochi, "Reusing recycled fibers in high-value fiber-reinforced polymer composites: Improving bending strength by surface cleaning," *Compos. Sci. Technol.*, vol. 72, no. 11, pp. 1298–1303, 2012, doi: http://dx.doi.org/10.1016/j.compscitech.2012.05.003.

[33]   R. Abdallah *et al.*, "A critical review on recycling composite waste using pyrolysis for sustainable development," *Energies*, vol. 14, no. 18, 2021, doi: 10.3390/en14185748.

[34]   K. W. Kim, J. S. Jeong, K. H. An, and B. J. Kim, "A Low Energy Recycling Technique of Carbon Fibers-Reinforced Epoxy Matrix Composites," *Ind. Eng. Chem. Res.*, vol. 58, no. 2, pp. 618–624, 2019, doi: 10.1021/acs.iecr.8b02554.

[35]   S. Y. Ye, A. Bounaceur, Y. Soudais, and R. Barna, "Parameter optimization of the steam thermolysis: A process to recover carbon fibers from polymer-matrix composites," *Waste and Biomass Valorization*, vol. 4, no. 1, pp. 73–86, 2013, doi: 10.1007/s12649-013-9220-4.

[36]   G. Grause, T. Mochizuki, T. Kameda, and T. Yoshioka, "Recovery of glass fibers from glass fiber reinforced plastics by pyrolysis," *J. Mater. Cycles Waste Manag.*, vol. 15, no. 2, pp. 122–128, 2013, doi: 10.1007/s10163-012-0101-x.

[37]   M. Blazsó, "Pyrolysis for recycling waste composites," in *Management, Recycling and Reuse of Waste Composites*, Goodship, V.,Ed., Sawston, Cambridge, UK: Woodhead Publishing, 2010, pp. 102–121.

[38]   Y. Zhou and K. Qiu, "A new technology for recycling materials from waste printed circuit boards," *J. Hazard. Mater.*, vol. 175, no. 1–3, pp. 823–828, 2010, doi: 10.1016/j.jhazmat.2009.10.083.

[39]   P. T. Williams, A. M. Cunliffe, and N. Jones, "Recovery of value-added products from the pyrolytic recycling of glass-fibre-reinforced composite plastic waste," *J. Energy Inst.*, vol. 78, no. 2, pp. 51–61, 2005.

[40]   M. S. Qureshi *et al.*, "Pyrolysis of plastic waste: Opportunities and challenges," *J. Anal. Appl. Pyrolysis*, vol. 152, no. February, 2020, doi: 10.1016/j.jaap.2020.104804.

[41]   H. Bel, H. Frej, L. Romain, D. Perrin, P. Ienny, and G. Pierre, "Recovery and reuse of carbon fibre and acrylic resin from thermoplastic composites used in marine application," *Resour., Conserv. Recycl.*, vol. 173, p. 105705, 2021, doi: 10.1016/j.resconrec.2021.105705.

[42]   S. Li, S. Sun, H. Liang, S. Zhong, and F. Yang, "Production and characterization of polypropylene composites filled with glass fibre recycled from pyrolysed waste printed circuit boards," *Environ. Technol.*, vol. 35, no. 21, pp. 2743–2751, 2014, doi: 10.1080/09593330.2014.920049.

[43]   N. Morita, Y. Kawabata, T. Wajima, A. T. Saito, and H. Nakagome, "Effect of the bromine-based flame retardant plastic pyrolysis of hydrotalcite," *MATEC Web Conf.*, vol. 62, pp. 0–4, 2016, doi: 10.1051/matecconf/20166205002.

[44] F. A. López, M. I. Martín, F. J. Alguacil, J. M. Rincón, T. A. Centeno, and M. Romero, "Thermolysis of fibreglass polyester composite and reutilisation of the glass fibre residue to obtain a glass – ceramic material," *J. Anal. Appl. Pyrolysis*, vol. 93, pp. 104–112, 2012, doi: 10.1016/j.jaap.2011.10.003.

[45] L. Long, S. Sun, S. Zhong, W. Dai, J. Liu, and W. Song, "Using vacuum pyrolysis and mechanical processing for recycling waste printed circuit boards," *J. Hazard. Mater.*, vol. 177, no. 1–3, pp. 626–632, 2010, doi: 10.1016/j.jhazmat.2009.12.078.

[46] J. A. Hiltz, "Pyrolysis-gas chromatography/mass spectrometry identification of styrene cross-linked polyester and vinyl ester resins," *J. Anal. Appl. Pyrolysis*, vol. 22, no. 1–2, pp. 113–128, 1991, doi: 10.1016/0165-2370(91)85011-U.

[47] T. M. Huynh, U. Armbruster, and A. Martin, *Phenolic Compounds – Natural Sources, Importance and Applications: Perspective on Co-feeding of Phenolic Compounds into Existing Refinery Units.* London, UK: IntechOpen, 2017.

[48] T. Groetsch *et al.*, "A modular LCA/LCC-modelling concept for evaluating material and process innovations in carbon fibre manufacturing," *Procedia CIRP*, vol. 98, pp. 529–534, 2021, doi: 10.1016/j.procir.2021.01.146.

[49] Y. S. Song, J. R. Youn, and T. G. Gutowski, "Life cycle energy analysis of fiber-reinforced composites," *Compos. Part A Appl. Sci. Manuf.*, vol. 40, pp. 1257–1265, 2009, doi: 10.1016/j.compositesa.2009.05.020.

[50] L. Giorgini *et al.*, "Recovery of Carbon Fibers from Cured and Uncured Carbon Fiber Reinforced Composites Wastes and Their Use as Feedstock for a New Composite Production," 2015, doi: 10.1002/pc.

[51] Y. Zhou, W. Wu, and K. Qiu, "Recovery of materials from waste printed circuit boards by vacuum pyrolysis and vacuum centrifugal separation," *Waste Manag.*, vol. 30, no. 11, pp. 2299–2304, 2010, doi: 10.1016/j.wasman.2010.06.012.

[52] M. A. Nahil and "Recycling of carbon fibre reinforced polymeric waste for the production of activated carbon fibres.," *J. Anal. Appl. Pyrolysis*, vol. 91, no. 1, pp. 67–75. 2011.

[53] A. Greco, A. Maffezzoli, and G. Buccoliero, "Thermal and chemical treatments of recycled carbon fibres for improved adhesion to polymeric matrix," 2012, doi: 10.1177/0021998312440133.

[54] A. M. Cunliffe, N. Jones, and P. T. Williams, "Recycling of fibre-reinforced polymeric waste by pyrolysis : thermo-gravimetric and bench-scale investigations," J. Anal. Appl. Pyrolysis., vol. 70, no. 2, pp. 315–338, 2003.

[55] K. Stoeffler, S. Andjelic, N. Legros, J. Roberge, and S. B. Schougaard, "Polyphenylene sulfide (PPS) composites reinforced with recycled carbon fiber," *Compos. Sci. Technol.*, vol. 84, pp. 65–71, 2013, doi: 10.1016/j.compscitech.2013.05.005.

[56] D. Ingenier and C. D. T. Tecnol, "Recycling of the Solid Residue Obtained from the Pyrolysis of Fiberglass Polyester Sheet Molding Compound," *Adv. Polym. Technol.*, vol. 28, no. 2, pp. 141–149, 2009, doi: 10.1002/adv.

[57] WindEurope, "Accelerating Wind Turbine Blade Circularity," *Themat. reports*, no. May, pp. 11–13, 2020.

[58] M. Gharfalkar, R. Court, C. Campbell, Z. Ali, and G. Hillier, "Analysis of waste hierarchy in the European waste directive 2008/98/EC," *Waste Manag.*, vol. 39, pp. 305–313, 2015, doi: 10.1016/j.wasman.2015.02.007.

[59] European Committee for Standardization, "EN15359: Solid Recovered Fuels: Specifications and Classes," vol. PTI/17, 2011.

[60] T. Hahn, "Holcim Deutschland Gruppe-Umweltdaten Bindemittel," 2017, https://scholar.google.com/scholar_lookup?title=Holcim%20Deutschland%20Gruppe-Umweltdaten%20Bindemittel&publication_year=2017&author=T.%20Hahn;https://www.holcim.de/sites/germany/files/atoms/files/holcim_umweltdaten_2017_web.pdf.

[61] J. Amanda, "Composites can be recycled," *Reinf. Plast.*, vol. 55, no. 3, pp. 45–46, 2011., doi: 10.1016/s0034-3617(11)70079-0.

[62] B. Taylor, "Veolia diverting GE windmill blades from landfill," *recycling today*, 2020, https://www.recy clingtoday.com/news/veolia-ge-windmill-blades-recycling-cement-kilns/.

[63] D. Kunii and O. Levenspiel, *Fluidization Engineering*, 2nd ed. Oxford, UK: Butterworth-Heinemann, 1991.

[64] N. Fenwick, "Recycling of composite materials using fluidised bed processes," no. October, 1996, [Online]. Available: http://etheses.nottingham.ac.uk/2837/.

[65] J. Kennerley and S. J. Pickering, "Recycling fibres recovered from composite materials using a fluidised bed process," University of Nottingham, 1998.

[66] K. Pender, "Recycling, regenerating and reusing reinforcement glass fibres." 2018, doi: 10.48730/ aa2e-c158.

[67] E. G. Melby and J. M. Castro, "Glass-reinforced Thermosetting Polyester Molding: Materials and Processing," *Compr. Polym. Sci. Suppl.*, pp. 51–109, 1989, doi: 10.1016/b978-0-08-096701-1.00206-8.

[68] F. Meng, "Environmental and cost analysis of carbon fibre composites recycling," The University of Nottingham, 2017.

[69] S. Melendi-Espina, C. N. Morris, T. A. Turner, and S. J. Pickering, "Recycling of carbon fibre composites," 2016, [Online]. Available: https://ueaeprints.uea.ac.uk/id/eprint/59606/1/ExtendedAb stract_S.Melendi_Espina.pdf.

[70] F. Meng, J. McKechnie, T. A. Turner, and S. J. Pickering, "Energy and environmental assessment and reuse of fluidised bed recycled carbon fibres," *Compos. Part A Appl. Sci. Manuf.*, vol. 100, pp. 206–214, 2017, doi: https://doi.org/10.1016/j.compositesa.2017.05.008.

[71] J. R. Kennerley, R. M. Kelly, N. J. Fenwick, S. J. Pickering, and C. D. Rudd, "The characterisation and reuse of glass fibres recycled from scrap composites by the action of a fluidised bed process," *Compos. Part A Appl. Sci. Manuf.*, vol. 29A, pp. 839–845, 1998, doi: 10.1016/S1359-835X(98)00008-6.

[72] H. Mason, "Moving toward next-generation wind blade recycling," *composites world*, 2022, https:// www.compositesworld.com/articles/moving-toward-next-generation-wind-blade-recycling.

[73] F. Meng, J. McKechnie, and S. J. Pickering, "An assessment of financial viability of recycled carbon fibre in automotive applications," *Compos. Part A Appl. Sci. Manuf.*, vol. 109, pp. 207–220, 2018, doi: https://doi.org/10.1016/j.compositesa.2018.03.011.

[74] W. Kaminsky, "Chemical recycling of plastics by fluidized bed pyrolysis," *Fuel Commun.*, vol. 8, no. May, p. 100023, 2021, doi: 10.1016/j.jfueco.2021.100023.

[75] F. Sasse and G. Emig, "Chemical recycling of polymer materials," *Chem. Eng. Technol.*, vol. 21, no. 10, pp. 777–789, 1998, doi: 10.1002/(SICI)1521-4125(199810)21:10<777::AID-CEAT777>3.0.CO;2-L.

[76] J. Ma, J. Wang, X. Tian, and H. Zhao, "In-situ gasification chemical looping combustion of plastic waste in a semi-continuously operated fluidized bed reactor," *Proc. Combust. Inst.*, vol. 37, no. 4, pp. 4389–4397, 2019, doi: 10.1016/j.proci.2018.07.032.

[77] F. J. Mastral, E. Esperanza, C. Berrueco, M. Juste, and J. Ceamanos, "Fluidized bed thermal degradation products of HDPE in an inert atmosphere and in air-nitrogen mixtures," *J. Anal. Appl. Pyrolysis*, vol. 70, no. 1, pp. 1–17, 2003, doi: 10.1016/S0165-2370(02)00068-2.

[78] K. Pender and L. Yang, "Investigation of the potential for catalysed thermal recycling in glass fibre reinforced polymer composites by using metal oxides," *Compos. Part A Appl. Sci. Manuf.*, vol. 100, pp. 285–293, 2017, doi: https://doi.org/10.1016/j.compositesa.2017.05.016.

[79] M. B. Gawande, R. K. Pandey, and R. V. Jayaram, "Role of mixed metal oxides in catalysis science-versatile applications in organic synthesis," *Catal. Sci. Technol.*, vol. 2, no. 6, pp. 1113–1125, 2012, doi: 10.1039/C2CY00490A.

[80] W. K. Lewis, E. R. Gilliland, and W. A. Reed, "Reaction of Methane with Copper Oxide in a Fluidized Bed," *Ind. Eng. Chem.*, vol. 41, no. 6, pp. 1227–1237, 1949, doi: 10.1021/ie50474a018.

[81] G. Pacchioni, "Oxygen Vacancy: The Invisible Agent on Oxide Surfaces," *ChemPhysChem*, vol. 4, no. 10, pp. 1041–1047, 2003, doi: 10.1002/cphc.200300835.

[82] J. Deng *et al.*, "Recycling of carbon fibers from CFRP waste by microwave thermolysis," *Processes*, vol. 7, no. 4, pp. 1–12, 2019, doi: 10.3390/pr7040207.

[83] Z. Peng *et al.*, "Dielectric characterization of Indonesian low-rank coal for microwave processing," *Fuel Process. Technol.*, vol. 156, pp. 171–177, 2017, doi: 10.1016/j.fuproc.2016.11.001.

[84] M. Oghbaei and O. Mirzaee, "Microwave versus conventional sintering: A review of fundamentals, advantages and applications," *J. Alloys Compd.*, vol. 494, no. 1–2, pp. 175–189, 2010, doi: 10.1016/j.jallcom.2010.01.068.

[85] D. El Khaled, N. Novas, J. A. Gazquez, and F. Manzano-Agugliaro, "Microwave dielectric heating: Applications on metals processing," *Renew. Sustain. Energy Rev.*, vol. 82, no. October 2017, pp. 2880–2892, 2018, doi: 10.1016/j.rser.2017.10.043.

[86] G. Oliveux, L. O. Dandy, and G. A. Leeke, "Current status of recycling of fibre reinforced polymers: Review of technologies, reuse and resulting properties," *Prog. Mater. Sci.*, vol. 72, pp. 61–99, 2015, doi: http://dx.doi.org/10.1016/j.pmatsci.2015.01.004.

[87] E. Lester, S. Kingman, K. H. Wong, C. Rudd, S. Pickering, and N. Hilal, "Microwave heating as a means for carbon fibre recovery from polymer composites: a technical feasibility study," *Mater. Res. Bull.*, vol. 39, no. 10, pp. 1549–1556, 2004, doi: http://dx.doi.org/10.1016/j.materresbull.2004.04.031.

[88] K. Obunai, T. Fukuta, and K. Ozaki, "Carbon fiber extraction from waste CFRP by microwave irradiation," *Compos. Part A Appl. Sci. Manuf.*, vol. 78, pp. 160–165, 2015, doi: 10.1016/j.compositesa.2015.08.012.

[89] D. Åkesson, Z. Foltynowicz, J. Christéen, and M. Skrifvars, "Microwave pyrolysis as a method of recycling glass fibre from used blades of wind turbines," *J. Reinf. Plast. Compos.*, vol. 31, no. 17, pp. 1136–1142, 2012, doi: 10.1177/0731684412453512.

[90] L. Jiang *et al.*, "Recycling carbon fiber composites using microwave irradiation: Reinforcement study of the recycled fiber in new composites," *J. Appl. Polym. Sci.*, vol. 132, no. 41, 2015, doi: 10.1002/app.42658.

[91] L. Yang, E. R. Sáez, U. Nagel, and J. L. Thomason, "Can thermally degraded glass fibre be regenerated for closed-loop recycling of thermosetting composites?," *Compos. Part A Appl. Sci. Manuf.*, vol. 72, pp. 167–174, 2015, doi: http://dx.doi.org/10.1016/j.compositesa.2015.01.030.

[92] J. L. Thomason, U. Nagel, L. Yang, and E. Sáez, "Regenerating the strength of thermally recycled glass fibres using hot sodium hydroxide," *Compos. Part A Appl. Sci. Manuf.*, vol. 87, pp. 220–227, 2016, doi: http://dx.doi.org/10.1016/j.compositesa.2016.05.003.

[93] S. Feih, E. Boiocchi, G. Mathys, Z. Mathys, A. G. Gibson, and A. P. Mouritz, "Mechanical properties of thermally-treated and recycled glass fibres," *Compos. Part B Eng.*, vol. 42, no. 3, pp. 350–358, 2011, doi: http://dx.doi.org/10.1016/j.compositesb.2010.12.020.

[94] P. G. Jenkins, L. Yang, J. J. Liggat, and J. L. Thomason, "Investigation of the strength loss of glass fibre after thermal conditioning," *J. Mater. Sci.*, vol. 50, no. 3, pp. 1050–1057, 2014, doi: 10.1007/s10853-014-8661-x.

[95] J. L. Thomason, C. C. Kao, J. Ure, and L. Yang, "The strength of glass fibre reinforcement after exposure to elevated composite processing temperatures," *J. Mater. Sci.*, vol. 49, no. 1, pp. 153–162, 2013, doi: 10.1007/s10853-013-7689-7.

[96] S. T. Bashir, L. Yang, R. Anderson, P. L. Tang, J. J. Liggat, and J. L. Thomason, "A simple chemical approach to regenerating the strength of thermally damaged glass fibre," *Compos. Part A Appl. Sci. Manuf.*, vol. 102, pp. 76–87, 2017, doi: http://dx.doi.org/10.1016/j.compositesa.2017.07.023.

[97] J. L. Thomason, L. Yang, and R. Meier, "The properties of glass fibres after conditioning at composite recycling temperatures," *Compos. Part A Appl. Sci. Manuf.*, vol. 61, pp. 201–208, 2014, doi: http://dx.doi.org/10.1016/j.compositesa.2014.03.001.

[98] J. R. Kennerley, N. J. Fenwick, S. J. Pickering, and C. D. Rudd, "The properties of glass fibers recycled from the thermal processing of scrap thermoset composites," *J. Vinyl Addit. Technol.*, vol. 3, no. 1, pp. 58–63, 1997, doi: 10.1002/vnl.10166.

[99] A. M. Cunliffe, N. Jones, and P. T. Williams, "Pyrolysis of composite plastic waste," *Environ. Technol.*, vol. 24, pp. 653–663, 2003.

[100] W. F. Thomas, "An investigation of the factors likely to affect the strength and properties of glass fibres," *Phys. Chem. Glas.*, vol. 1, no. 1, pp. 4–18, 1960.

[101] P. J. James Thomason Liu Yang, "Glass Fibre Strength-A Review with Relation to Composite Recycling," *Fibers*, vol. 4, p. 18, 2016, doi: doi:10.3390/fib4020018.

[102] S. Feih, A. P. Mouritz, and S. W. Case, "Determining the mechanism controlling glass fibre strength loss during thermal recycling of waste composites," *Compos. Part A Appl. Sci. Manuf.*, vol. 76, pp. 255–261, 2015, doi: http://dx.doi.org/10.1016/j.compositesa.2015.06.006.

[103] P. K. Gupta, "Strength of Glass Fibers," in *Fiber Fracture*, Elices, M., Llorca, J., Eds., Amsterdam, Netherlands: Elsevier Science, 2002, pp. 127–153.

[104] S. Sakka, "Effects of reheating on strength of glass fibres," *Bull. Inst. Chem. Res. Kyoto Univ.*, vol. 34, no. 6, pp. 316–320, 1957.

[105] W. H. Otto, "Compaction Effects in Glass Fibers," *J. Am. Ceram. Soc.*, vol. 44, no. 2, pp. 68–72, 1961, doi: 10.1111/j.1151-2916.1961.tb15352.x.

[106] L. Yang and J. L. Thomason, "The thermal behaviour of glass fibre investigated by thermomechanical analysis," *J. Mater. Sci.*, vol. 48, no. 17, pp. 5768–5775, 2013, doi: 10.1007/s10853-013-7369-7.

[107] A. Fraisse, J. Beauson, P. Brodsted, and B. Madsen, "Thermal recycling and re-manufacturing of glass fibre thermosetting composites," *IOP Conf. Ser. Mater. Sci. Eng.*, vol. 139, no. 012020, 2016, doi: 10.1088/1757-899X/139/1/012020.

[108] R. S. Ginder and S. Ozcan, "Recycling of Commercial E-glass Reinforced Thermoset Composites via Two Temperature Step Pyrolysis to Improve Recovered Fiber Tensile Strength and Failure Strain," *Recycling*, vol. 4, no. 24, pp. 24, 2019.

[109] C. C. Kao, J. L. Thomason, A. C. Group, and U. Kingdom, "Regeneration of Thermally Recycled Glass Fibre for Cost-Effective Composite Recycling : Performance of Fibre Recyclates from Thermoset Composites and with Subsequent Recover Treatments," in *ECCM16 – 16th European Conference on Composite Materials*, 2014, no. June, pp. 22–26.

[110] S. Feih and A. P. Mouritz, "Tensile properties of carbon fibres and carbon fibre-polymer composites in fire," *Compos. Part A Appl. Sci. Manuf.*, vol. 43, no. 5, pp. 765–772, 2012, doi: 10.1016/j.compositesa.2011.06.016.

[111] G. Jiang and S. J. Pickering, "Structure–property relationship of recycled carbon fibres revealed by pyrolysis recycling process," *J. Mater. Sci.*, vol. 51, no. 4, pp. 1949–1958, 2016, doi: 10.1007/s10853-015-9502-2.

[112] H. L. H. Yip, S. J. Pickering, and C. D. Rudd, "Characterisation of carbon fibres recycled from scrap composites using fluidised bed process," *Plast. Rubber Compos.*, vol. 31, no. 6, pp. 278–282, 2002, doi: 10.1179/146580102225003047.

[113] S. J. Pickering *et al.*, "Developments in the fluidised bed process for fibre recovery from thermoset composites," 2015, [Online]. Available: https://www.researchgate.net/publication/301849557_Developments_in_the_fluidised_bed_process_for_fibre_recovery_from_thermoset_composites/link/5854164608ae77ec370458a2/download?_tp=eyJjb250ZXh0Ijp7ImZpcnN0UGFnZSI6InB1YmxpY2F0aW9uW9uIiwicGFnZSI6InB1YmxpY2F0aW9uW9aW.

[114] S. Hao, L. He, J. Liu, Y. Liu, C. Rudd, and X. Liu, "Recovery of carbon fibre from waste prepreg via microwave pyrolysis," *Polymers (Basel).*, vol. 13, no. 8, pp. 1–15, 2021, doi: 10.3390/polym13081231.

[115] J. S. Jeong, K. W. Kim, K. H. An, and B. J. Kim, "Fast recovery process of carbon fibers from waste carbon fibers-reinforced thermoset plastics," *J. Environ. Manage.*, vol. 247, no. July, pp. 816–821, 2019, doi: 10.1016/j.jenvman.2019.07.002.

[116] J. L. Thomason and L. J. Adzima, "Sizing up the interphase: an insider's guide to the science of sizing," *Compos. Part A Appl. Sci. Manuf.*, vol. 32, no. 3–4, pp. 313–321, 2001, doi: http://dx.doi.org/10.1016/S1359-835X(00)00124-X.

[117] U. Nagel, L. Yang, C. C. Kao, and J. L. Thomason, "Effects of Thermal Recycling Temperatures on the Reinforcement Potential of Glass Fibers," *Polym. Compos.*, 2016, doi: 10.1002/pc.24029.

[118] S. Rudzinski, L. Häussler, C. Harnisch, E. Mäder, and G. Heinrich, "Glass fibre reinforced polyamide composites: Thermal behaviour of sizings," *Compos. Part A Appl. Sci. Manuf.*, vol. 42, no. 2, pp. 157–164, 2011, doi: http://dx.doi.org/10.1016/j.compositesa.2010.10.018.

[119] P. Gao, Y. Ward, K. B. Su, and L. T. Weng, "Effects of chemical composition and thermal stability of finishes on the compatibility between glass fiber and high melting temperature thermoplastics," *Polym. Compos.*, vol. 21, no. 2, pp. 312–321, 2000, doi: 10.1002/pc.10188.

[120] Q.-T. Pham and C.-S. Chern, "Thermal stability of organofunctional polysiloxanes," *Thermochim. Acta*, vol. 565, pp. 114–123, 2013, doi: http://dx.doi.org/10.1016/j.tca.2013.04.032.

[121] C. Roux, J. Denault, and M. F. Champagne, "Parameters regulating interfacial and mechanical properties of short glass fiber reinforced polypropylene," *J. Appl. Polym. Sci.*, vol. 78, no. 12, pp. 2047–2060, 2000, doi: 10.1002/1097-4628(20001213)78:12<2047::aid-app10>3.0.co;2-z.

[122] D. Bikiaris, P. Matzinos, A. Larena, V. Flaris, and C. Panayiotou, "Use of silane agents and poly (propylene-*g*-maleic anhydride) copolymer as adhesion promoters in glass fiber/polypropylene composites," *J. Appl. Polym. Sci.*, vol. 81, no. 3, pp. 701–709, 2001, doi: 10.1002/app.1487.

[123] U. Zielke, K. J. Hüttinger, and W. P. Hoffman, "Surface-oxidized carbon fibers: I. Surface structure and chemistry," *Carbon N. Y.*, vol. 34, no. 8, pp. 983–998, 1996, doi: 10.1016/0008-6223(96)00032-2.

[124] M. Wada, K. Kawai, T. Suzuki, H. Hira, and S. Kitaoka, "Effect of superheated steam treatment of carbon fiber on interfacial adhesion to epoxy resin," *Compos. Part A Appl. Sci. Manuf.*, vol. 85, pp. 156–162, 2016, doi: 10.1016/j.compositesa.2016.03.024.

[125] G. Jiang, S. J. Pickering, G. S. Walker, K. H. Wong, and C. D. Rudd, "Surface characterisation of carbon fibre recycled using fluidised bed," *Appl. Surf. Sci.*, vol. 254, no. 9, pp. 2588–2593, 2008, doi: 10.1016/j.apsusc.2007.09.105.

[126] G. Cai, M. Wada, I. Ohsawa, S. Kitaoka, and J. Takahashi, "Interfacial adhesion of recycled carbon fibers to polypropylene resin: Effect of superheated steam on the surface chemical state of carbon fiber," *Compos. Part A Appl. Sci. Manuf.*, vol. 120, pp. 33–40, 2019, doi: 10.1016/j.compositesa.2019.02.020.

[127] A. Fernández, M. Santangelo-Muro, J. P. Fernández-Blázquez, C. S. Lopes, and J. M. Molina-Aldareguia, "Processing and properties of long recycled-carbon-fibre reinforced polypropylene," *Compos. Part B Eng.*, vol. 211, no. October 2020, p. 108653, 2021, doi: 10.1016/j.compositesb.2021.108653.

[128] J. Howarth and F. R. Jones, "Interface optimisation of recycled carbon fibre composites," *ECCM 2012 – Compos. Venice, Proc. 15th Eur. Conf. Compos. Mater.*, no. June, pp. 24–28, 2012.

[129] A. Salas et al., "Ultrafast carbon nanotubes growth on recycled carbon fibers and their evaluation on interfacial shear strength in reinforced composites," *Sci. Rep.*, vol. 11, no. 1, pp. 1–11, 2021, doi: 10.1038/s41598-021-84419-y.

[130] K. H. Wong, D. Syed Mohammed, S. J. Pickering, and R. Brooks, "Effect of coupling agents on reinforcing potential of recycled carbon fibre for polypropylene composite," *Compos. Sci. Technol.*, vol. 72, no. 7, pp. 835–844, 2012, doi: 10.1016/j.compscitech.2012.02.013.

[131] D. T. Burn et al., "The usability of recycled carbon fibres in short fibre thermoplastics : interfacial properties," *J. Mater. Sci.*, vol. 51, no. 16, pp. 7699–7715, 2016, doi: 10.1007/s10853-016-0053-y.

[132] I. De marco et al., "Recycling of the Products Obtained in the Pyrolysis of Fibre-Glass Polyester SMC," *J. Chem. Technol. Biotechnol.*, vol. 69, no. 2, pp. 187–192, 1997, doi: 10.1002/(sici)1097-4660(199706)69:2<187::aid-jctb710>3.0.co;2-t.

[133] N.-J. Lee and J. Jang, "The use of a mixed coupling agent system to improve the performance of polypropylene-based composites reinforced with short-glass-fibre mat," *Compos. Sci. Technol.*, vol. 57, no. 12, pp. 1559–1569, 1998, doi: http://dx.doi.org/10.1016/S0266-3538(97)00086-9.

[134] S. J. Pickering, T. A. Turner, and N. A. Warrior, "Moulding compound development using recycled carbon fibres," In This is proceeding from the 38th International SAMPE Technical Conference (ISTC), Dallas, Texas, 2006.

[135] N. Feng, X. Wang, and D. Wu, "Surface modification of recycled carbon fiber and its reinforcement effect on nylon 6 composites: Mechanical properties, morphology and crystallization behaviors," *Curr. Appl. Phys.*, vol. 13, no. 9, pp. 2038–2050, 2013, doi: 10.1016/j.cap.2013.09.009.

[136] H. Han, X. Wang, and D. Wu, "Preparation, crystallization behaviors, and mechanical properties of biodegradable composites based on poly(L-lactic acid) and recycled carbon fiber," *Compos. Part A Appl. Sci. Manuf.*, vol. 43, no. 11, pp. 1947–1958, 2012, doi: 10.1016/j.compositesa.2012.06.014.

[137] G. Yan, X. Wang, and D. Wu, "Development of lightweight thermoplastic composites based on polycarbonate/acrylonitrile-butadiene-styrene copolymer alloys and recycled carbon fiber: Preparation, morphology, and properties," *J. Appl. Polym. Sci.*, vol. 129, no. 6, pp. 3502–3511, 2013, doi: 10.1002/app.39105.

[138] H. Han, X. Wang, and D. Wu, "Mechanical properties, morphology and crystallization kinetic studies of bio-based thermoplastic composites of poly(butylene succinate) with recycled carbon fiber," *J. Chem. Technol. Biotechnol.*, vol. 88, no. 7, pp. 1200–1211, 2013, doi: 10.1002/jctb.3956.

[139] Y. Chen, X. Wang, and D. Wu, "Recycled carbon fiber reinforced poly(butylene terephthalate) thermoplastic composites: Fabrication, crystallization behaviors and performance evaluation," *Polym. Adv. Technol.*, vol. 24, no. 4, pp. 364–375, 2013, doi: 10.1002/pat.3088.

[140] F. Manis, G. Stegschuster, J. Wölling, and S. Schlichter, "Influences on textile and mechanical properties of recycled carbon fiber nonwovens produced by carding," *J. Compos. Sci.*, vol. 5, no. 8, pp. 1–16, 2021, doi: 10.3390/jcs5080209.

[141] K. Giannadakis, M. Szpieg, and J. Varna, "Mechanical Performance of a Recycled Carbon Fibre/PP Composite," *Exp. Mech.*, vol. 51, no. 5, pp. 767–777, 2011, doi: 10.1007/s11340-010-9369-8.

[142] J. Wölling, M. Schmieg, F. Manis, and K. Drechsler, "Nonwovens from Recycled Carbon Fibres – Comparison of Processing Technologies," *Procedia CIRP*, vol. 66, pp. 271–276, 2017, doi: 10.1016/j.procir.2017.03.281.

[143] M. Szpieg, M. Wysocki, and L. E. Asp, "Reuse of polymer materials and carbon fibres in novel engineering composite materials," *Plast. Rubber Compos.*, vol. 38, no. 9–10, pp. 419–425, 2009, doi: 10.1179/146580109X12540995045688.

[144] K. H. Wong, T. A. Turner, and S. J. Pickering, "Challenges in developing nylon composites comingled with discontinuous recycled carbon fibre," *16th Eur. Conf. Compos. Mater. ECCM 2014*, no. June, pp. 22–26, 2014.

[145] D. U. Shah and P. J. Schubel, "On recycled carbon fibre composites manufactured through a liquid composite moulding process," *J. Reinf. Plast. Compos.*, vol. 35, no. 7, pp. 533–540, 2016, doi: 10.1177/0731684415623652.

[146] M. L. Longana, N. Ong, H. Yu, and K. D. Potter, "Multiple closed loop recycling of carbon fibre composites with the HiPerDiF (High Performance Discontinuous Fibre) method," *Compos. Struct.*, vol. 153, no. Supplement C, pp. 271–277, 2016, doi: https://doi.org/10.1016/j.compstruct.2016.06.018.

[147] D. Heider *et al.*, "Carbon Fiber Composites Recycling Technology Enabled by the TuFF Technology," *SAMPE J.*, 2022, https://www.nasampe.org/store/viewproduct.aspx?id=21240459.

[148] S. Pimenta and S. T. Pinho, "The effect of recycling on the mechanical response of carbon fibres and their composites," *Compos. Struct.*, vol. 94, no. 12, pp. 3669–3684, 2012, doi: 10.1016/j.compstruct.2012.05.024.

[149] W. H. Bowyer and M. G. Bader, "On the re-inforcement of thermoplastics by imperfectly aligned discontinuous fibres," *J. Mater. Sci.*, vol. 7, no. 11, pp. 1315–1321, 1972, doi: 10.1007/bf00550698.

[150] J. L. Thomason, "The influence of fibre length and concentration on the properties of glass fibre reinforced polypropylene. 6. The properties of injection moulded long fibre PP at high fibre content," *Compos. Part A Appl. Sci. Manuf.*, vol. 36, no. 7, pp. 995–1003, 2005, doi: http://dx.doi.org/10.1016/j.compositesa.2004.11.004.

[151] N.-J. Lee and J. Jang, "The effect of fibre content on the mechanical properties of glass fibre mat/ polypropylene composites," *Compos. Part A Appl. Sci. Manuf.*, vol. 30, no. 6, pp. 815–822, 1999, doi: http://dx.doi.org/10.1016/S1359-835X(98)00185-7.

[152] J. L. Thomason and M. A. Vlug, "Influence of fibre length and concentration on the properties of glass fibre-reinforced polypropylene: 1. Tensile and flexural modulus," *Compos. Part A Appl. Sci. Manuf.*, vol. 27, no. 6, pp. 477–484, 1996, doi: http://dx.doi.org/10.1016/1359-835X(95)00065-A.

[153] J. L. Thomason, "The influence of fibre length and concentration on the properties of glass fibre reinforced polypropylene: 5. Injection moulded long and short fibre PP," *Compos. Part A Appl. Sci. Manuf.*, vol. 33, no. 12, pp. 1641–1652, 2002, doi: http://dx.doi.org/10.1016/S1359-835X(02)00179-3.

[154] D. E. Spahr, K. Friedrich, J. M. Schultz, and R. S. Bailey, "Microstructure and fracture behaviour of short and long fibre-reinforced polypropylene composites," *J. Mater. Sci.*, vol. 25, no. 10, pp. 4427–4439, 1990, doi: 10.1007/bf00581104.

[155] R. K. Mittal and V. B. Gupta, "The strength of the fibre-polymer interface in short glass fibre-reinforced polypropylene," *J. Mater. Sci.*, vol. 17, no. 11, pp. 3179–3188, 1982, doi: 10.1007/bf01203481.

[156] G. D. Tomkinson-Walles, "Performance of Random Glass Mat Reinforced Thermoplastics," *J. Thermoplast. Compos. Mater.*, vol. 1, no. 1, pp. 94–106, 1988, doi: 10.1177/089270578800100108.

[157] U. Nagel and P. D. J. Thomason, "The processing and characterisation of recycled glass fibre composites," University of Strathclyde, UK, 2016.

[158] J. G. Iglesias, J. González-Benito, A. J. Aznar, J. Bravo, and J. Baselga, "Effect of Glass Fiber Surface Treatments on Mechanical Strength of Epoxy Based Composite Materials," *J. Colloid Interface Sci.*, vol. 250, no. 1, pp. 251–260, 2002, doi: https://doi.org/10.1006/jcis.2002.8332.

[159] X. M. Liu, J. L. Thomason, and F. R. Jones, "The Concentration of Hydroxyl Groups on Glass Surfaces and Their Effect on the Structure of Silane Deposits," in *Silanes and Other Coupling Agents, Volume 5*, Mittal, K. L., Eds., Leiden, Netherlands: Brill Academic Publishers, CRC Press, 2009, pp. 25–38.

[160] J. González-Benito, J. Baselga, and A. J. Aznar, "Microstructural and wettability study of surface pretreated glass fibres," *J. Mater. Process. Technol.*, vol. 92, pp. 129–134, 1999, doi: http://dx.doi.org/10.1016/S0924-0136(99)00212-5.

[161] S. K. Gopalraj and T. Kärki, "A Study to Investigate the Mechanical Properties of Recycled Carbon Fibre/Glass Fibre-Reinforced Epoxy," *Processes*, vol. 8, no. 9, pp. 954, https://doi.org/10.3390/pr8080954.

[162] C. Bowland, "formulation study of long fiber thermoplastic polypropylene (Part 2): The effects of coupling agent type and properties," *Automot. Compos. Conf. Expo.* 2009.

[163] T. Knudsen, J. Bech, and K. Elvebakken, "Forbehandling af hærdeplastbaserede kompositmaterialer til genanvendelse," Danish Environmental Protection Agency, no. 4, 2005, https://www2.mst.dk/udgiv/publikationer/2005/87-7614-605-7/pdf/87-7614-606-5.pdf.

[164] T. A. Turner, N. A. Warrior, and S. J. Pickering, "Development of high value moulding compounds from recycled carbon fibres," *Plast. Rubber Compos.*, vol. 39, no. 3–5, pp. 151–156, 2010, doi: 10.1179/174328910X12647080902295.

[165] J. Shi, L. Bao, R. Kobayashi, J. Kato, and K. Kemmochi, "Reusing recycled fibers in high-value fiber-reinforced polymer composites: Improving bending strength by surface cleaning," *Compos. Sci. Technol.*, vol. 72, no. 11, pp. 1298–1303, 2012, doi: 10.1016/j.compscitech.2012.05.003.

[166] P. R. Wilson, A. Ratner, G. Stocker, F. Syred, K. Kirwan, and S. R. Coles, "Interlayer hybridization of virgin carbon, recycled carbon and natural fiber laminates," *Materials (Basel).*, vol. 13, no. 21, pp. 1–16, 2020, doi: 10.3390/ma13214955.

[167] S. Pimenta, S. T. Pinho, P. Robinson, K. H. Wong, and S. J. Pickering, "Mechanical analysis and toughening mechanisms of a multiphase recycled CFRP," *Compos. Sci. Technol.*, vol. 70, no. 12, pp. 1713–1725, 2010, doi: 10.1016/j.compscitech.2010.06.017.

[168] M. L. Longana, V. Ondra, H. Yu, K. D. Potter, and I. Hamerton, "Reclaimed carbon and flax fibre composites: Manufacturing and mechanical properties," *Recycling*, vol. 3, no. 4, 2018, doi: 10.3390/recycling3040052.

[169] B. Tse, X. Yu, H. Gong, and C. Soutis, "Flexural Properties of Wet-Laid Hybrid Nonwoven Recycled Carbon and Flax Fibre Composites in Poly-Lactic Acid Matrix," *Aerospace*, vol. 5, no. 4, p. 120, 2018, doi: 10.3390/aerospace5040120.

[170] M. Oliveira, K. L. Pickering, and C. Gauss, "Hybrid Polyethylene Composites with Recycled Carbon Fibres and Hemp Fibres Produced by Rotational Moulding," *J. Compos. Sci.*, vol. 6, p. 352, 2022.

[171] N. Shah, J. Fehrenbach, and C. A. Ulven, "Hybridization of hemp fiber and recycled-carbon fiber in polypropylene composites," *Sustain.*, vol. 11, no. 11, 2019, doi: 10.3390/su11113163.

[172] J. Shi, L. Bao, and K. Kemmochi, "Low-Velocity Impact Response and Compression After Impact Assessment of Recycled Carbon Fiber-Reinforced Polymer Composites for Future Applications," *Polym. Compos.*, vol. 35, pp. 1494–1506, 2013.

[173] "Fibreglass Recycling Europe – fibreglass recycling Europe." https://www.fibreglass-recycling.eu/france/ (accessed Jun. 02, 2022).

[174] FLSmidth, "Alternative fuels." https://www.flsmidth.com/en-gb/solutions/alternative-fuels.

[175] A. Durakovic, "DecomBlades Consortium Launches Blade Recycling Project," *offshoreWIND.biz*, 2021, https://www.offshorewind.biz/2021/01/25/decomblades-consortium-launches-blade-recycling-project/.

[176] H. Mason, "DecomBlades consortium awarded funding for a cross-sector wind turbine blade recycling project," *Composite World*, 2021, https://www.compositesworld.com/news/decomblades-consortium-awarded-funding-for-a-cross-sector-wind-turbine-blade-recycling-project.

[177] Makeen Energy, "Pyrolysing more than plastic – Achieving circularity for composites manufacturers." https://www.makeenenergy.com/products-solutions/innovations/composite-waste-conversion.

[178] Veolia and GE Renewable Energy, "Press release: GE Renewable Energy Announces US Blade Recycling Contract with Veolia," 2020, https://www.ge.com/news/press-releases/ge-renewable-energy-announces-us-blade-recycling-contract-with-veolia?utm_campaign=Veolia+recycling&utm_medium=bitly&utm_source=external-LI-Jerome.

[179] G. Nehls, "GE announces U.S. blade recycling contract with Veolia," *composites world*, 2020, https://www.compositesworld.com/news/ge-announces-us-blade-recycling-contract-with-veolia.

[180] G. Nehls, "Carbon Rivers to commercialize fiberglass composites upcycling project," *composites world*, 2021 https://www.compositesworld.com/news/carbon-rivers-to-commercialize-fiberglass-composites-upcycling-project.

[181] bcircular, "Collection of the wind blade in Bon Vent Vilalba wind farm." https://www.bcircular.com/collection-of-the-wind-blade-in-bon-vent-vilalba-wind-farm/.

[182] K. Larsen, "Recycling wind," *Reinforced Plastics*, vol. 53, no.1, pp. 20–23, 25, 2009.

[183] Vattenfall, "Press release: Turning wind turbines. . . into skis, insulation and solar farms," 2022. https://group.vattenfall.com/uk/newsroom/pressreleases/2022/turning-wind-turbines. . .-into-skis-insulation-and-solar-farms.

[184] C. Zhang, "Recycled Carbon Fiber From Rotor Blades Finds a New Life In Skiing," *Following the Wind*, 2022. http://following-the-wind.com/2022/02/18/recycled-carbon-fiber-from-rotor-blades-finds-a-new-life-in-skiing/.

[185] G. Nehls, "U.K. wind turbine blade recycling project, PRoGrESS, commences," *composites world*, 2022. https://www.compositesworld.com/news/uk-wind-turbine-blade-recycling-project-progress-commences (accessed Jan. 12, 2024).

[186] S. Black, "Composites recycling is gaining traction," *Composites World*, 2017. https://www.compositesworld.com/articles/composites-recycling-gaining-traction (accessed Dec. 20, 2024).

[187] "Reclaim, Reuse and Reap the Rewards," *Composites Manufacturing*, 2020. From "Composites Manufacturing" the official magazine of the "American Composites Manufacturing Association", https://acmanet.org/reclaim-reuse-and-reap-the-rewards/.

[188] "Hexcel invests in carbon fiber recycling company," *recycling today*, 2016, https://www.recyclingto day.com/news/hexcel-carbon-conversions-recycling/.

[189] G. Nehls, "Mitsubishi Chemical plans to acquire European carbon fiber recycling companies," *Composites World*, 2020. https://www.compositesworld.com/news/mitsubishi-chemical-plans-to-acquire-european-carbon-fiber-recycling-companies-.

[190] H. Mason, "Pressurized steam-based composites recycling for full fiber reclamation," *Composite World*, 2022, https://www.compositesworld.com/articles/pressurized-steam-based-composites-recy cling-for-full-fiber-reclamation.

[191] S. Francis, "The state of recycled carbon fiber," *Composite World*, 2019, https://www.composites world.com/articles/the-state-of-recycled-carbon-fiber.

[192] G. Nehls, "University of Delaware TuFF composite material shows high potential for UAM," *Composite World*, 2021, https://www.compositesworld.com/news/university-of-delaware-tuff-compos ite-material-shows-high-potential-for-uam.

[193] G. Gardiner, "Revolutionizing the composites cost paradigm, Part 1: Feedstock," *Composite World*, 2020, https://www.compositesworld.com/articles/revolutionizing-the-composites-cost-paradigm-part-1-feedstock.

[194] "carboNXT products," *Mitsubishi Chemical Advanced Materials*. https://www.carbonxt.de/en/products/.

Fanran Meng
4 Life cycle assessment for composite sustainability

4.1 Introduction

Life cycle assessment (LCA) serves as a standardised method for evaluating the environmental performance of products and materials across their entire life span, encompassing raw material extraction, manufacturing, transportation, use, and end-of-life (EoL) treatment [1, 2]. By quantifying energy and material flows, emissions, and waste outputs, LCA facilitates comparative assessments and supports environmentally conscious decision-making. When combined with economic analysis, LCA offers a comprehensive view of sustainability trade-offs, aiding the development of resource-efficient systems [3–6].

4.1.1 Methodology of LCA

The ISO 14040/44 framework outlines four essential steps for conducting LCA as shown in Figure 4.1 [7, 8]:
1) Goal and scope definition: Clearly outline the product or system under study, the functional unit, and the intended application of the assessment. The system boundary is also established here.
2) Inventory analysis (LCI): Compile data on all relevant energy and material inputs, outputs, and emissions.
3) Impact assessment (LCIA): Translate LCI data into potential environmental impacts across categories such as climate change, acidification, and eutrophication.
4) Interpretation: Synthesise results in line with study objectives to identify environmental hotspots or trade-offs.

During the goal and scope definition, the LCA must specify the product or process being evaluated and the intended application of the results. This includes setting a consistent functional unit for comparison and clearly outlining system boundaries to determine which activities and flows are included.

The inventory analysis builds the system model based on these boundaries. It requires gathering data across all processes such as material extraction, production,

https://doi.org/10.1515/9783110754438-004

energy use, and transportation. LCA tools such as ecoinven,[1] US LCI Database,[2] EU/JRC ELCD database,[3] BioEnergieDat,[4] GREET,[5] and GaBi[6] support inventory compilation.

LCIA aggregates inventory data into predefined impact categories, including climate change, resource depletion, human health, and ecosystem quality. Various methodologies exist for impact assessment (e.g. CML 2002, ReCiPe, Eco-Indicator 99, and IPCC), each with distinct category weightings. Software like SimaPro, GaBi, and EuCIA enables detailed LCIA and facilitates interpretation.

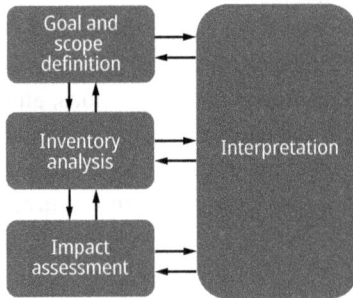

Figure 4.1: The typical LCA structure.

4.1.2 LCA in the context of composites

Thermoset or thermoplastic matrices combined with glass or carbon fibres are increasingly employed in sectors prioritising weight reduction and performance. LCA enables quantification of environmental benefits and drawbacks when replacing traditional materials (e.g. steel) with fibre-reinforced composites in transport and energy systems.

Figure 4.2 illustrates the life cycle of carbon fibre-reinforced plastic (CFRP) and glass fibre-reinforced plastic (GFRP), encompassing raw material acquisition, component fabrication, usage, and waste management. Although some LCAs overlook the EoL stage [3, 9, 10], recent research has included recycling, incineration, and landfill impacts [11–16]. For instance, Witik et al. [4] assessed CFRP substitution for steel in automotive applications and found marginal benefits due to the energy-intensive nature of virgin CF (vCF) production. As illustrated in Figure 4.3, the lifetime CO_2 reduction benefits of using CF materials to replace steel are strongly influenced by the assumed vehicle travel distance. Longer travel distances lead to greater total fuel

---

1 http://www.ecoinvent.org/.
2 http://www.nrel.gov/lci/.
3 http://eplca.jrc.ec.europa.eu/ELCD3/.
4 http://www.bioenergiedat.de/.
5 https://greet.anl.gov/net.
6 http://www.gabi-software.com/uk-ireland/index/.

savings due to the lightweight nature of CF components. Nevertheless, all studies consistently highlight that CF production is highly energy-intensive and results in substantially higher greenhouse gas (GHG) emissions compared to conventional materials [17–18].

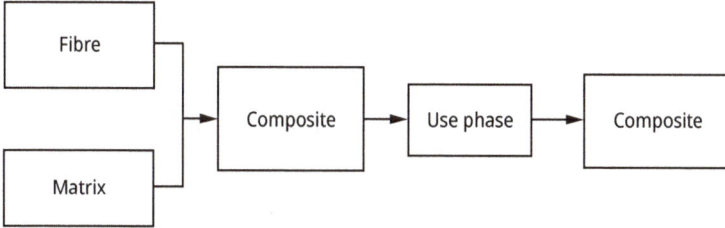

Figure 4.2: Life cycle stages of composite materials.

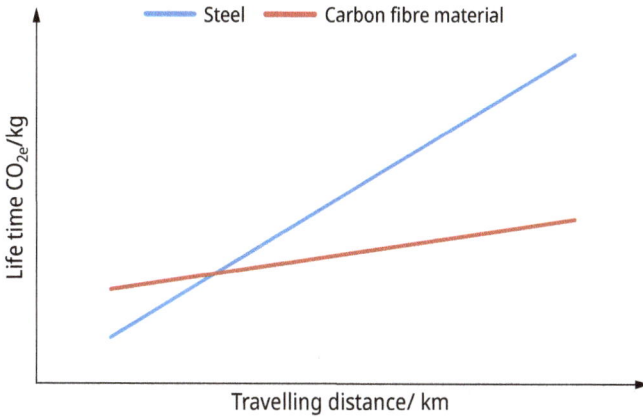

Figure 4.3: Lifetime CO_2 emissions to the travelling distance of a vehicle using steel and CF materials, respectively.

4.2 Life cycle inventory of carbon fibre

4.2.1 Production of carbon fibre

Polyacrylonitrile (PAN) is the dominant precursor for CF, accounting for around 90% of production [19, 20]. Alternative precursors (e.g. biomass-derived lignin [21]) are under development but not fully commercialised. The manufacturing process involves polymerisation, fibre spinning, stabilisation (oxidation), carbonisation, surface treatment, and sizing. During stabilisation, fibres are oxidised in air at about 230–280 °C, followed by high-temperature carbonisation in inert environments up to 1,700 °C. Graphitisation, where applicable, occurs at even higher temperatures to enhance mechanical performance. Figure 4.4 outlines the full production pathway.

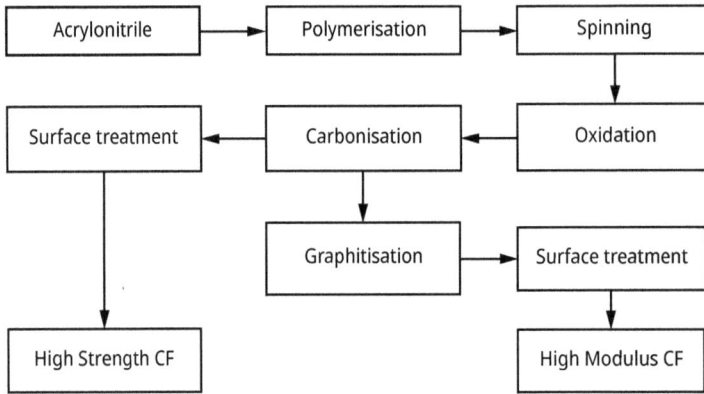

Figure 4.4: PAN-based carbon fibre production process.

The transformation involves significant energy use and results in substantial weight loss (up to 45%) as volatile gases are released [22]. The final product properties (e.g. modulus and strength) depend on processing parameters like heat treatment levels.

4.2.2 Inventory data and variability

Reliable LCI data for CF production remains limited due to commercial confidentiality. Reported energy use varies widely (198–595 MJ/kg), reflecting differences in assumptions, data sources, and process configurations [9, 21, 23–37]. Table 4.1 compiles data from literature and industry sources, revealing substantial inconsistencies.

Table 4.1: Reported energy use for carbon fibre production.

Direct energy consumed (MJ/kg CF)	Reference	Origin
22.7	[23]	Calculated
478	[9, 24]	Original data from a producer
171	[25]	Original data
478, 286	[10], JCMA, 2006 [26]	JCMA, METI (Ministry of Economy, Trade and Industry)-Industrial data
400	[27]	Personal communication
198–595	[28]	Original data from a producer

Table 4.1 (continued)

Direct energy consumed (MJ/kg CF)	Reference	Origin
353	[29]	Original data
183–286	[30]	Previous publication [10]
9.62	[31]	Calculated
405.24	[21]	Original data from a producer
478,286	[32]	JCMA
198–594	[33]	Previous publication [28]
353	[34, 35]	Previous publication [29]
9.62	[36]	Previous publication [31]
353	[37]	Previous publication [29]

Studies often fail to relate energy consumption to fibre properties or account for process-specific parameters. For example, Duflou et al. [29] and Das [21] used different assumptions for energy inputs, yielding divergent estimates. Table 4.2 compares key parameters in both datasets.

Table 4.2: Comparison of Duflou and Das datasets for CF manufacture.

Parameters	Energy use	Energy mix	Yield	Non-energy Inputs
Duflou	162 MJ electricity and 191 MJ natural gas, 33.87 kg steam	Electricity, steam, natural gas	53%	AN, nitrogen, DGEBA
Das	75 MJ electricity, 330.24 MJ natural gas	Electricity, natural gas	45.6%	AN, vinyl acetate, solvent

In addition to energy use, emissions from CF production (e.g. CO_2, HCN, and NH_3) are critical for LCA but are rarely quantified in detail. Only a few studies attempt to estimate emissions based on stoichiometric calculations [31, 38]. These gaps highlight the need for standardised, disaggregated inventory models based on real industrial operations.

Emerging approaches like lignin-derived CF or microwave-assisted carbonisation promise lower cost and energy use. US Oak Ridge National Laboratory estimates up to 50% cost savings and 60% energy reduction through alternative precursors and novel technologies [39–41].

4.2.3 Sensitivity analysis

LCI data for vCF production remains limited and inconsistent. Reported energy demands range widely – from 198 to 595 MJ/kg – due to variations in electricity, natural
gas, and steam use [10, 28, 29, 43]. Furthermore, existing studies fail to correlate these
inputs with fibre properties despite the differing processing conditions needed to produce high-modulus or high-strength vCF. In this chapter, a reference case assumes
149.4 MJ electricity, 177.8 MJ natural gas, and 31.4 kg steam per kg of vCF. Direct emissions are estimated based on available literature [31, 42]. Sensitivity analysis includes
low (198 MJ/kg) and high (595 MJ/kg) energy cases, using the same energy mix ratios
as the reference (see shaded area in Figure 4.5).

The carbon footprint of vCF production is highly dependent on the electricity
source. For instance, using hydroelectric power (7 $gCO_2eq./kWh$) results in only 29 kg
$CO_2eq./kg$ vCF, compared to 68 kg $CO_2eq./kg$ when coal-based electricity (960 $gCO_2eq./$
kWh) is used (Figure 4.5). Under the highest energy use scenario, GHG emissions from
rCF are just 9% of those from vCF versus 5% under the UK grid mix. Due to vCF's high
energy intensity, manufacturers are increasingly turning to renewable energy – such
as BMW and SGL's hydropower-based production in Moses Lake, USA – for both environmental and cost benefits. These results also reinforce the value of rCF, balancing
emissions savings with the challenge of maintaining equivalent functional performance to vCF.

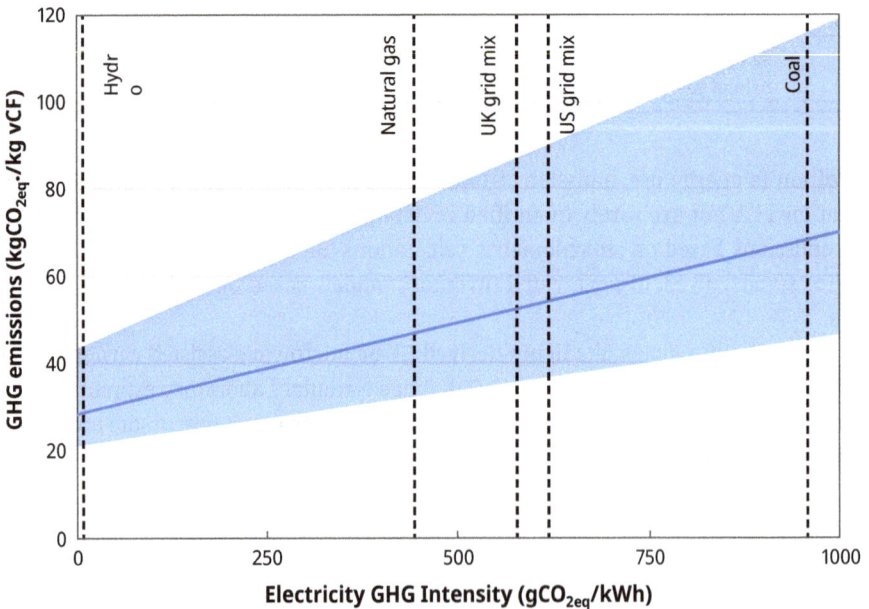

Figure 4.5: Impact of electricity source on GHG emissions of vCF production.

4.3 LCA of composite manufacturing

4.3.1 Matrix materials

The matrix phase in fibre-reinforced composites plays a crucial role in binding the reinforcement and transferring loads. Both thermosets (e.g. epoxy and polyester) and thermoplastics (e.g. polypropylene and polyethylene) are widely used. The energy required for resin production varies significantly due to differences in feedstocks and chemical processing technologies. For example, epoxy resin is derived from petroleum-based precursors and involves multi-step synthesis routes that are energy-intensive, typically requiring between 76 and 137 MJ/kg. Comparatively, polypropylene, a thermoplastic, may consume as little as 24 MJ/kg under optimised conditions.

Material choice is dictated by the intended application: thermosets are favoured for structural performance and thermal stability, especially in aerospace and automotive sectors, while thermoplastics allow for reshaping and recycling, making them increasingly attractive in circular economy models. Table 4.3 presents energy intensities of typical polymer matrices used in composites.

Table 4.3: Energy intensity of common resin systems.

Matrix	Energy intensity (MJ/kg)	References
Epoxy resin	76–137	[10, 30, 44–46]
Unsaturated polyester	62.8–78	[10, 30, 45, 46]
Phenol	32.9	[10, 45, 46]
Flexible polyurethane	67.3	[10, 45, 46]
High-density polyethylene	20.3	[10, 45, 46]
Low-density polyethylene	65–92	[30, 45, 46]
Polypropylene	24.4–112	[10, 30, 45–47]
Polyvinyl chloride	53–80	[30, 45, 46]
Polystyrene	71–118	[30, 45, 46]

4.3.2 Manufacturing processes

Composite manufacturing involves forming techniques tailored to the type of matrix and reinforcement. Techniques such as resin transfer moulding, hand lay-up, autoclave curing, pultrusion, and filament winding have unique energy demands and operational efficiencies. Autoclaving, although effective for high-performance parts, requires elevated pressure and temperature, leading to energy intensities as high as 135 MJ/kg. In contrast, pultrusion and filament winding, used for continuous profile production, consume less than 5 MJ/kg.

Process selection must consider part geometry, production volume, and performance targets. Additionally, energy consumption can vary based on curing time, pressure levels, tooling, and automation. Table 4.4 presents energy intensities of common manufacturing methods used in composites production.

Table 4.4: Energy requirements of manufacturing methods [3, 10, 21, 30, 47, 48].

Manufacturing methods	Energy intensity (MJ/kg)
Spray up	14.9
Filament winding	2.7
Hand lay-up	19.2
Pultrusion	3.1
Resin transfer moulding	12.8
Injection moulding (hydraulic)	19
Vacuum-assisted resin infusion	10.2
Sheet moulding compound	3.5
Cold press	11.8
Preform matched die	10.1
Prepreg production	40
Autoclave moulding	21.9–135
Compression moulding	9.06

4.4 LCA of composite recycling

End-of-life strategies for CFRP waste are increasingly important due to growing concerns over landfill use and incineration impacts. Recycling presents a promising approach, offering the potential to recover the value embedded in composite materials while mitigating the environmental burdens of vCF production. Effective recycling methods must maintain sufficient fibre integrity to enable reuse, and their environmental and financial performance should be assessed holistically through LCA.

Current recycling technologies for CFRP span mechanical, thermal (e.g. pyrolysis and fluidised bed), and chemical routes. Despite increasing interest, few studies have systematically evaluated the life cycle implications of these processes. For instance, Li et al. [49] examined mechanical recycling and found that it can lower GHG emissions, primary energy demand, and landfill volumes compared to conventional disposal. This benefit stems from its relatively low energy intensity – estimated at 0.27–2.03 MJ/kg for CFRP (at 10–150 kg/h throughput) [14, 50]. However, the method was not cost-competitive when replacing virgin glass fibre due to high operational costs and limited material recovery value.

Thermal recycling methods, especially pyrolysis, are still emerging. Estimated energy demand ranges from 3 MJ/kg for GFRP to 30 MJ/kg for CFRP [50]. However, life cycle data for these technologies are sparse. Witik et al. [34] evaluated pyrolysis against landfill and incineration but relied on speculative energy values, limiting reliability. Financial feasibility was also not addressed.

Chemical depolymerisation techniques offer another route. Shibata and Nakagawa [51] investigated rCF production via solvent and catalyst-based depolymerisation from CFRP tennis rackets (50 wt.% CF). Energy use declined with scale: 91 MJ/kg at 1,000 rackets/month, down to 63 MJ/kg at 17,000 rackets/month, with distillation alone consuming up to 38 MJ/kg. This highlights the need to optimise solvent recovery to lower energy demand.

Keith et al. [52] measured energy for a solvolysis process (excluding solvent recovery) and found an energy intensity of 19.2 MJ/kg rCF. Lower processing temperatures and increased reactor throughput were suggested to reduce energy consumption. While solvolysis can recover both fibres and chemicals, solvent and product recovery steps can increase environmental burdens.

The fluidised bed process involves oxidising the epoxy matrix at about 500 °C in a sand bed reactor. Released fibres are separated and gas products are combusted in a secondary chamber for energy recovery. Previous studies [53] modelled the FB process at 500 t/year plant capacity and 9 kg/m^2·hr feed rate, reporting energy demands of 1.9 MJ/kg (natural gas) and 1.7 kWh/kg (electricity). Stoichiometric equations based on a resin mixture of 87% DGEBA and 13% IPD estimate CO_2 and NO_2 emissions from matrix oxidation. Depending on scenario assumptions, recovered CF from FB recycling uses only 1–20% of the energy required for virgin CF production. Figure 4.6 summarises the suite of composite recycling options explored in this section.

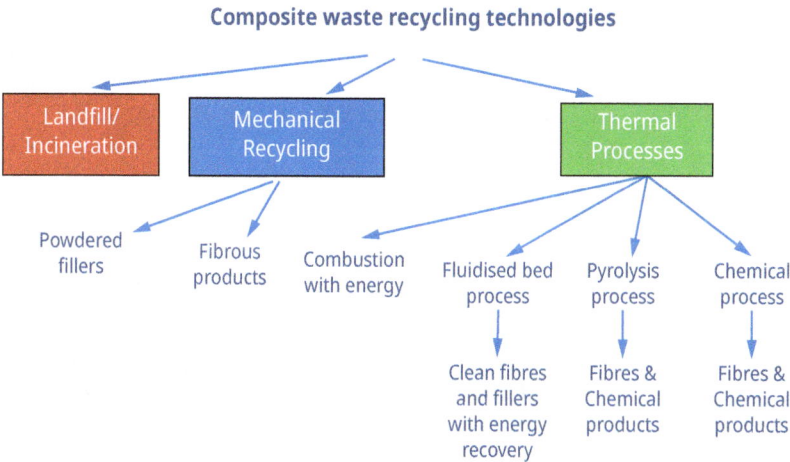

Figure 4.6: Composite recycling technologies evaluated in this chapter.

4.4.1 Landfilling

Sanitary landfilling of composite waste is considered a terminal option with no further emissions or energy use post-disposal [46, 49]. However, it does not enable material recovery and contributes to long-term waste accumulation.

4.4.2 Municipal incineration

Incineration can recover energy embedded in composites. Assuming 13% electricity generation efficiency and 38% combined heat and power efficiency, outputs offset regional electricity grids and natural gas heating. CO_2 emissions resulting from the oxidation of the epoxy matrix material are calculated on a stoichiometric basis assuming all carbon is fully oxidised to CO_2 and all nitrogen is emitted as NO_2 (see eqs. (4.1) and (4.2)). Around 60 wt.% of residue is sent to landfill after incineration:

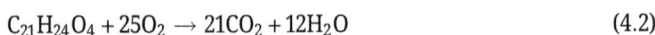

$$C_{10}H_{22}N_2 + 17.5O_2 \rightarrow 10CO_2 + 11H_2O + 2NO_2 \tag{4.1}$$

$$C_{21}H_{24}O_4 + 25O_2 \rightarrow 21CO_2 + 12H_2O \tag{4.2}$$

4.4.3 Cement kiln co-processing

Composite waste can be incinerated in cement kilns, displacing petroleum coke (35.8 MJ/kg) with a minor efficiency loss (3%) [54, 55]. This approach provides energy recovery and material substitution in a high-temperature industrial process.

4.4.4 Mechanical recycling

Mechanical recycling breaks down composite waste into two fractions: 42% coarse/fine fibres and 30% powder [56, 57]. The fibres can replace virgin glass fibre at a substitution ratio of 0.78 due to property degradation [57, 58], while the powder can substitute calcium carbonate in cement kilns. Remaining coarse residues (28%) are landfilled or incinerated. Figure 4.7 illustrates the process.

4.4.5 Pyrolysis recycling

Pyrolysis involves heating shredded composites in the absence of oxygen (300–800 °C), producing recycled fibres and hydrocarbon by-products. Fibres retain ~ 52% of tensile strength, and the matrix is partly recovered as liquid fuels [59, 60]. ELG Carbon Fibre Ltd (now called Gen 2 Carbon) provides commercial data on energy inputs [62].

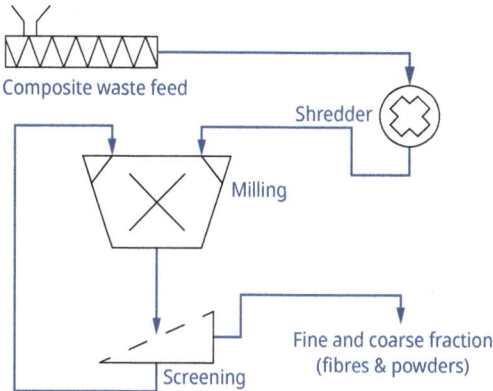

Figure 4.7: Mechanical recycling process overview.

Most lab-scale setups do not recover pyrolysis gases or vapours due to cost and technical challenges, but future commercial systems could improve recovery efficiency. Figure 4.8 depicts the pyrolysis process.

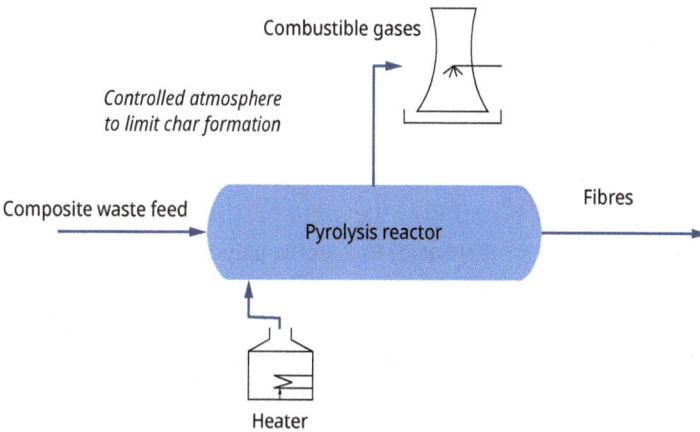

Figure 4.8: Pyrolysis recycling process for composites.

4.4.6 Fluidised bed recycling

In a fluidised bed process, composite waste is pre-shredded and oxidised at about 500 °C in a fluidised silica sand bed. Combustion gases are oxidised at 750 °C and passed through heat exchangers and boilers. Four fans maintain airflow through the system. Total energy use is ~ 3.3 MJ/kg, including 1.1 MJ of natural gas and 2.1 MJ of electricity [42, 62, 63]. Fibre strength loss ranges from 10 to 50% depending on temperature and fibre length [64, 65]. Figure 4.9 outlines the FB process.

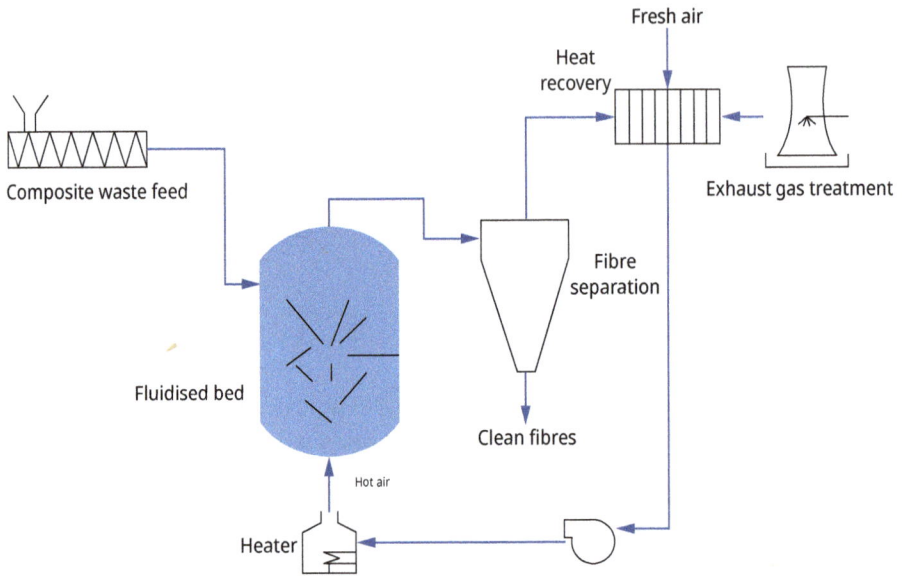

Figure 4.9: Schematic of fluidised bed recycling system.

4.4.7 Chemical recycling

Chemical recycling uses solvents (e.g. nitric acid) to dissolve polymer matrices and recover fibres and chemicals. Recovered fibres typically lose 33–53% of tensile strength [66]. Phenol is the dominant degradation product, with others including isopropyl phenol and cresol. Assumed recovery per kilogram of epoxy composite waste includes 0.29 kg phenol, 0.08 kg trimethyl benzene, and 0.03 kg aniline [67]. Figure 4.10 illustrates the chemical recycling setup.

4.4.8 Life cycle inventory

The LCI phase compiles detailed data on material and energy flows during recycling and composite production. Process simulations are validated with experimental and pilot-scale data. Peer-reviewed literature and commercial LCA databases (e.g. ecoinvent and GaBi) fill data gaps and provide industry benchmarks. A robust LCI underpins accurate environmental assessment of composite recycling technologies.

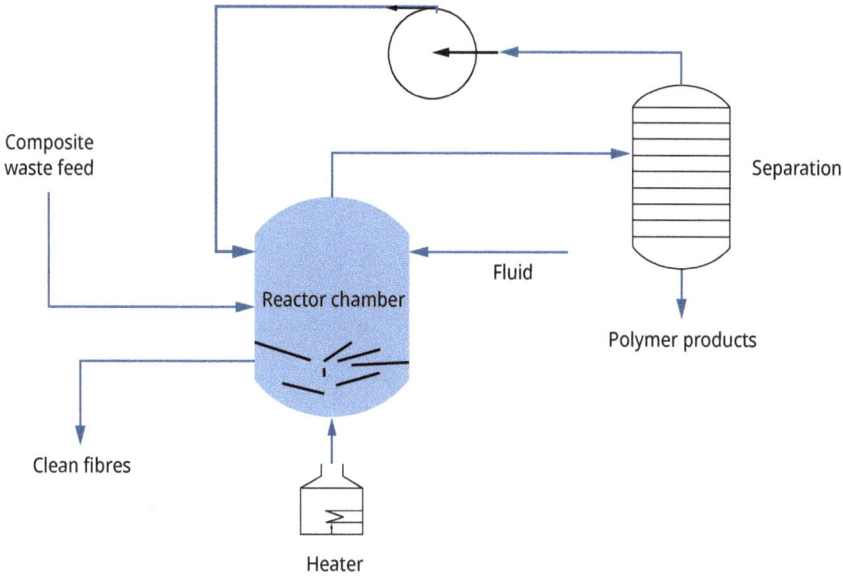

Figure 4.10: Chemical recycling via solvent decomposition.

4.5 Life cycle assessment of CFRP manufacture from rCF

When compared to virgin CFRP, recycled CFRP (rCFRP) offers substantial reductions in environmental impact. For compression-moulded components, the total primary energy requirement is approximately 55.5% (for equivalent stiffness) and 56.3% (for equivalent strength) relative to virgin CFRP (Figure 4.11a). This difference stems from the mechanical performance of rCFRP – where strength is more significantly affected than stiffness – requiring thicker composite panels to achieve equivalent functionality (7.5% thicker for strength, versus 5.2% for stiffness).

vCF production remains the most energy-intensive stage of the virgin CFRP life cycle, accounting for 47% of total energy demand. Polyamide synthesis adds another 22%, followed by compression moulding (24%) and wet papermaking (7%). In contrast, rCFRP benefits from the low energy use of the fluidised bed recycling process, which contributes just 2% to total energy consumption. Instead, polyamide production (41%) and compression moulding (44%) dominate energy use in the recycled route, with wet papermaking at 13% for equivalent stiffness panels.

Regarding climate impact, the global warming potential (GWP) for each compression-moulded composite part is 6.6 kgCO$_2$eq. for virgin CFRP, 3.4 kgCO$_2$eq. for rCFRP

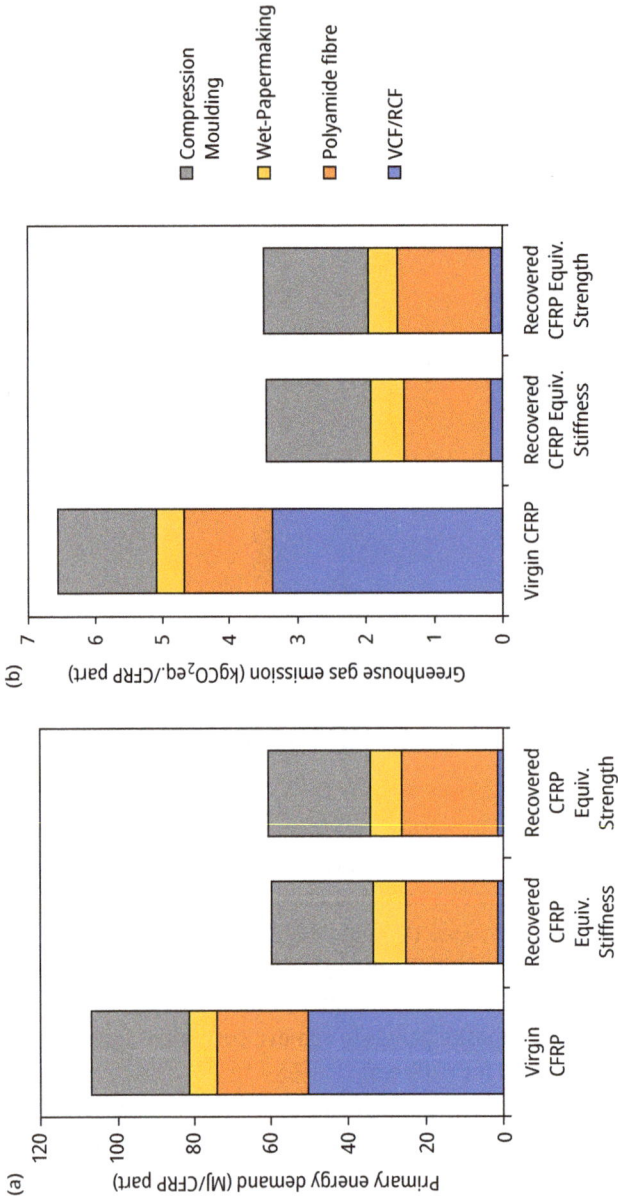

Figure 4.11: (a) Primary energy demand and (b) greenhouse gas emission comparisons for compression moulded composites produced from vCF and rCF.

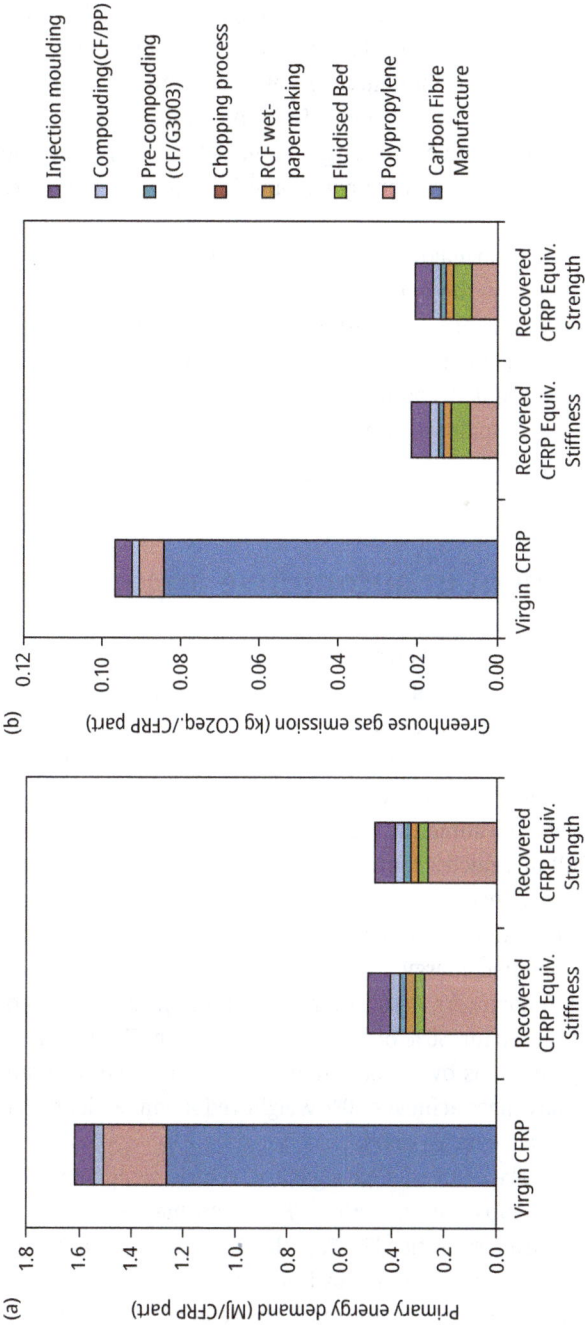

Figure 4.12: (a) Primary energy demand and (b) greenhouse gas emission comparisons for injection moulded composites produced from vCF and rCF.

with equivalent stiffness, and 3.5 kgCO$_2$eq. for equivalent strength (Figure 4.11b). Over half (52%) of the GHG emissions for virgin CFRP originate from vCF production.

For injection-moulded parts, rCFRP again demonstrates lower impacts as shown in Figure 4.12: primary energy use drops to 30% and 29% of the virgin CFRP case for equivalent stiffness and strength, respectively. In virgin composites, vCF production accounts for 78% of total energy demand. Polypropylene and injection moulding are also energy-intensive contributors.

In the recycled pathway, wet papermaking (13%), injection moulding (16%), and two-stage compounding (12%) comprise the majority of energy use. The fluidised bed recycling stage uses only 7% of total energy. Correspondingly, GWP for virgin injection-moulded parts is 96.6 kgCO$_2$eq. compared to 22.8 kgCO$_2$eq. and 22.3 kgCO$_2$eq. for rCFRP with the equivalent stiffness and strength, respectively. vCF accounts for 87% of virgin composite emissions, while polypropylene contributes 6.5–6.7 kgCO$_2$eq. in both scenarios.

4.6 Life cycle assessment in automotive applications

The use phase in automotive applications typically contributes 60–70% of total life cycle energy consumption [68]. Thus, material substitution with lighter alternatives – like CFRP or aluminium – can offer significant fuel and emissions savings over a vehicle's lifespan. Reported fuel savings due to lightweighting range from 0.15 to 0.48 L/ (100 km · 100 kg) [4, 69–72]. However, production-related energy demands must be carefully considered to assess net environmental benefits.

Several LCA studies of CFRP in vehicles show mixed results. While CFRP can reduce operational energy use, its production – particularly vCF – is extremely energy-intensive. Witik et al. [4] found only limited net benefits when replacing steel components with vCFRP due to high embodied energy.

The Japan Carbon Fibre Manufacturers Association (JCMA) analysed CFRP adoption across sectors. In aircraft, replacing 50% of body materials with CFRP reduces total mass by 20%, cutting CO$_2$ emissions by 27,000 tonnes over 10 years per aircraft. In cars, using CFRP for 17% of body mass achieves 30% weight reduction, with an estimated CO$_2$ saving of 5 ton per vehicle over 10 years.

Here, life cycle GHG emissions are evaluated for a generic vehicle component, initially made of mild steel. Material substitutions (vCF, rCF) are normalised for thickness and mass using mechanical property ratios [73–75]. The thickness ratio accounts for differences in stiffness between reference and substitute materials.

For a plate or flat panel, the ratio of component thickness required to account for varying material properties can be determined as

$$R_t = \frac{t}{t_{\text{ref}}} = \left(\frac{E_{\text{ref}}}{E}\right)^{\frac{1}{3}} \tag{4.3}$$

where R_t is the thickness ratio between the proposed lightweight material (t) and the reference (mild steel, t_{ref}) and E is the modulus of the two materials (GPa).

Material properties of vCFRP and random rCFRP are obtained from experiments [76], manufacturers [77, 78], and properties of aligned rCFRP re derived from micromechanical models [79, 80]. Additional reference properties (steel, aluminium, magnesium) are sourced from databases [17, 81, 82]. Table 4.5 presents material properties and corresponding relative thicknesses of component materials.

Table 4.5: Material properties of selected engineering materials.

Material	Matrix	Manufacture	Density (g/cm³)	Modulus (GPa)	Strength (MPa)	Refs
Mild steel	–	Stamping	7.81	207.00	350.00	[82]
Random rCF 30%	Epoxy resin	Compression moulding	1.38	37.14	314.44	[77]
Woven vCF 50%	Epoxy resin	Autoclave moulding	1.60	70.00	570.00	[79]

During use, added vehicle mass increases fuel consumption. In-use energy is estimated using the EPA Physical Emission Rate Estimator [83] and mathematical models [84], applied to a representative mid-size sedan (Ford Fusion). A 200,000 km vehicle lifetime is assumed.

Results show that while woven vCFRP yields the lightest components, high production emissions can offset in-use savings. Conversely, rCFRP offers significant life cycle reductions in energy, emissions, and cost (Figure 4.13). GHG performance is sensitive to electricity mix as shown in Figure 4.14: hydroelectric-powered production lowers emissions by 35% (woven vCF, aligned rCFRP) and 20% (random rCFRP) versus the UK grid. Ongoing electricity decarbonisation will further enhance lightweight materials' benefits.

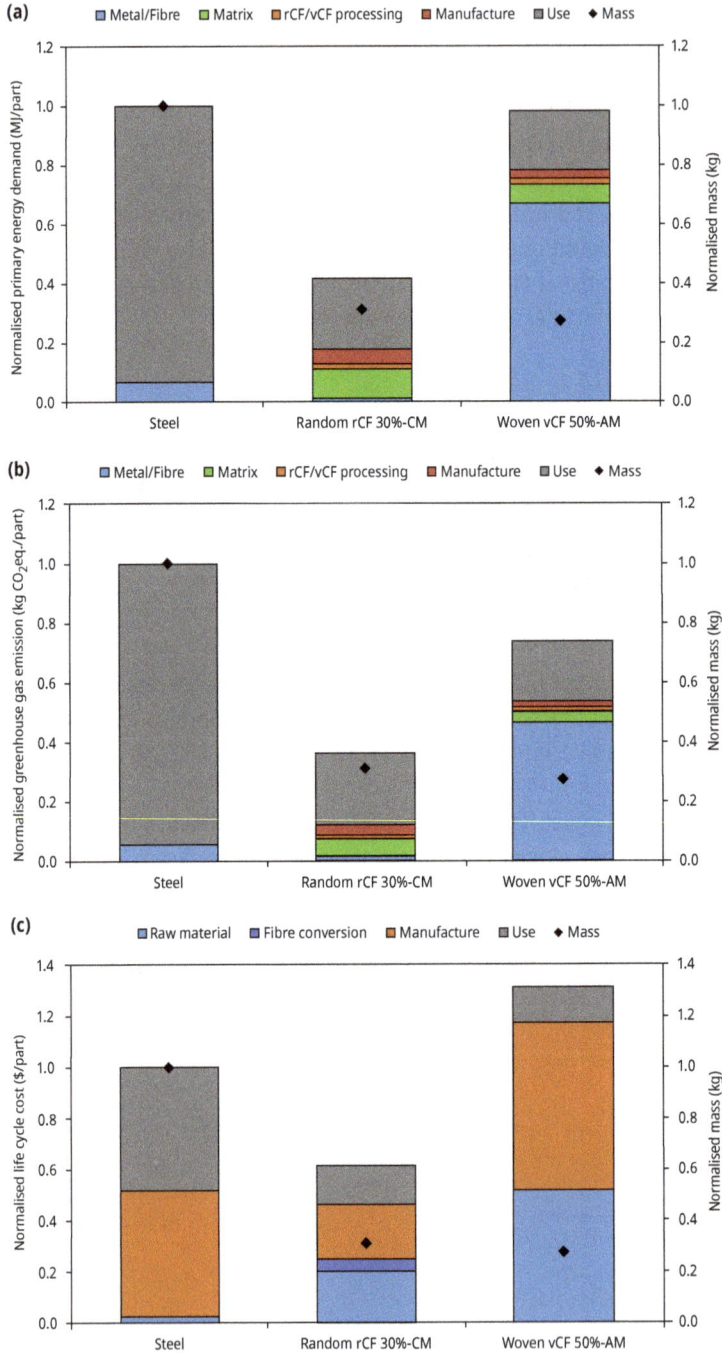

Figure 4.13: Life cycle (a) primary energy, (b) GHG emissions, and (c) cost and mass of functionally equivalent automotive components.

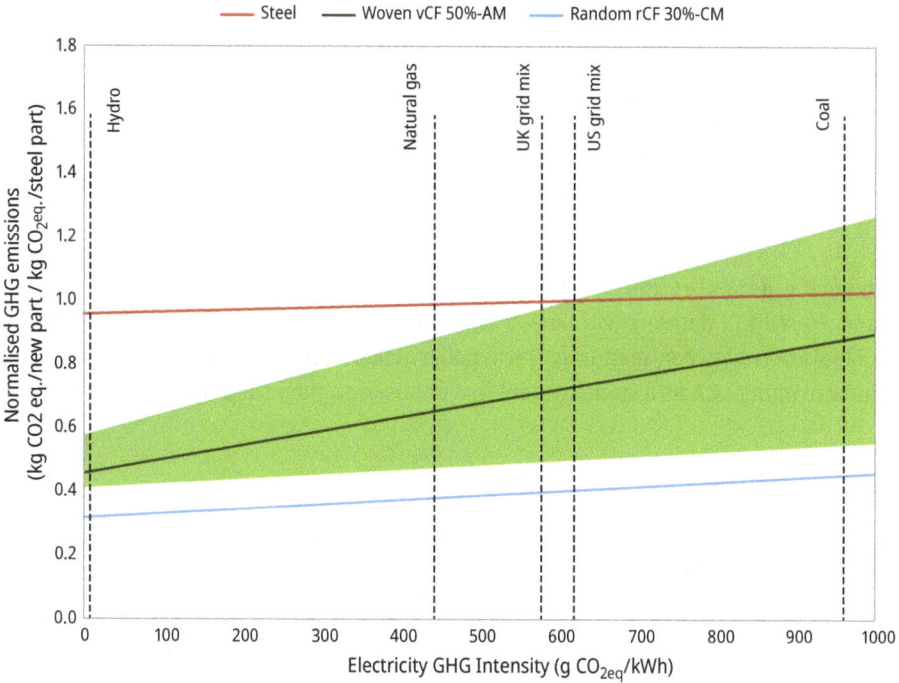

Figure 4.14: Sensitivity of GHG emissions to grid electricity intensity and vCF energy input uncertainty.

4.7 Conclusions

LCA is a crucial tool for evaluating the environmental impact of composite materials across their full life cycle from raw material extraction to disposal or recycling. This chapter quantified the primary energy use and GHG emissions of virgin and recycled CF composites, demonstrating the significant environmental benefits of using rCF in automotive applications. Beyond reducing environmental burdens, rCF enhances the economic feasibility of recycling technologies and supports circular economy initiatives.

Wider adoption of rCF could drive sustainability in transportation and other industries. As policymakers and manufacturers develop circular strategies, LCA provides critical insights to inform material selection, promote recycling, and improve supply chain transparency. Future research should refine LCA models with region- and process-specific data, particularly for CF manufacturing, while collaboration among researchers, industry, and policymakers is essential for establishing robust LCI datasets. Equipping engineers and scientists with LCA expertise will further ensure sustainable materials innovation in the transition to a low-carbon economy.

Key findings from this chapter include:
- Virgin CF production is highly energy-intensive, with significant variability across sources.
- End-of-life recycling methods reduce GHG emissions and energy use, though maintaining fibre quality remains a challenge.
- Integrating LCA into composite design supports data-driven material choices and circular economy pathways.

Advancing data transparency, improving recycling technologies, and developing standardised LCI datasets will be essential for achieving sustainability in composites. Collaboration between manufacturers, policymakers, and researchers will be key to mainstreaming LCA as a standard tool in composite engineering.

References

[1]   O'Neill TJ. Life Cycle Assessment and Environmental Impact of Polymeric Products: iSmithers Rapra Publishing; Shrewsbury, Shropshire, UK: Rapra Technology Limited, 2003.

[2]   Henrikke Bumann A-MT. The Hitch Hiker's Guide to LCA. Lund, Sweden: Studentlitteratur AB; 2004.

[3]   Witik RA, Gaille F, Teuscher R, Ringwald H, Michaud V, Månson J-AE. Economic and environmental assessment of alternative production methods for composite aircraft components. J Clean Prod. 2012;29–30(0): 91–102.

[4]   Witik RA, Payet J, Michaud V, Ludwig C, Månson J-AE. Assessing the life cycle costs and environmental performance of lightweight materials in automobile applications. Compos Part A-Appl S. 2011;42(11):1694–1709.

[5]   Schwab Castella P, Blanc I, Gomez Ferrer M, Ecabert B, Wakeman M, Manson J-A, et al. Integrating life cycle costs and environmental impacts of composite rail car-bodies for a Korean train. Int J Life Cycle Ass. 2009;14(5):429–442.

[6]   Ilg P, Hoehne C, Guenther E. High-performance materials in infrastructure: a review of applied life cycle costing and its drivers – the case of fiber-reinforced composites. J Clean Prod. 2016;112:926–945.

[7]   14040 I. Environmental Management: Life Cycle Assessment: Principles and Framework, Geneva: International Standards Organization. 2006a.

[8]   ISO. Environmental Management: Life Cycle Assessment: Requirements and Guidelines, Geneva: International Standards Organization. 2006b.

[9]   Nagai H, Takahashi J, Kemmochi K, Matsui J-i, Sakai S. Inventory analysis of energy consumption on advanced polymer-based composite materials. Journal of the National Institute of Materials and Chemical Research. 2000;8(4):161–169.

[10]   Suzuki T, Takahashi J. Prediction of energy intensity of carbon fiber reinforced plastics for mass-produced passenger cars. The Ninth Japan International SAMPE symposium 2005. p. 14–19.

[11]   Longana ML, Ong N, Yu H, Potter KD. Multiple closed loop recycling of carbon fibre composites with the HiPerDiF (High Performance Discontinuous Fibre) method. Compos Struct. 2016;153:271–277.

[12]   La Rosa AD, Banatao DR, Pastine SJ, Latteri A, Cicala G. Recycling treatment of carbon fibre/epoxy composites: Materials recovery and characterization and environmental impacts through life cycle assessment. Composites Part B: Engineering. 2016;104:17–25.

[13] Keith MJ, Oliveux G, Leeke GA. Optimisation of solvolysis for recycling carbon fibre reinforced composites. In Proceedings of the ECCM17—17th European Conference on Composite Materials, Munich, Germany, 26–30 June 2016.

[14] Howarth J, Mareddy SSR, Mativenga PT. Energy intensity and environmental analysis of mechanical recycling of carbon fibre composite. J Clean Prod. 2014;81(0):46–50.

[15] Jiang G, Wong W, Pickering S, Rudd C, Walker G. Study of a fluidised bed process for recycling carbon fibre from polymer composite. 7th world congress for chemical engineering, Glasgow, UK 2005.

[16] Pickering SJ. Recycling technologies for thermoset composite materials—current status. Compos Part A-Appl S. 2006;37(8):1206–1215.

[17] Kelly JC, Sullivan JL, Burnham A, Elgowainy A. Impacts of Vehicle Weight Reduction via Material Substitution on Life-Cycle Greenhouse Gas Emissions. Environ Sci Technol. 2015;49(20):12535–12542.

[18] Suzuki T, Teshiba F, Zu SH, Takahashi J, Kageyama K, Yoshinari H. Life Cycle Assessment of Lightweight Automobiles using CFRP. JSME Annual Meeting: The Japan Society of Mechanical Engineers; 2002. p. 281–282, https://www.jstage.jst.go.jp/article/jsmemecjo/2002.2/0/2002.2_281/_article/-char/ja/.

[19] Zoltek. Carbon fibre: How is it made? 2017 http://zoltek.com/carbonfiber/how-is-it-made/; https://toray-cfe.com/en/what-is-carbon-fiber/.

[20] The Japan Carbon Fiber Manufacturers Association. 2016.

[21] Das S. Life cycle assessment of carbon fiber-reinforced polymer composites. The International Journal of Life Cycle Assessment. 2011;16(3):268–282.

[22] Delhaes P. Fibers and Composites: Delhaes, P., Ed. Fibers and Composites. No. 22632. New York: Taylor & Francis,; 2003.

[23] Lee SM, Jonas T, Disalvo G. The beneficial energy and environmental-impact of composite-materials – an unexpected bonus SAMPE Journal. 1991;27(2):19–25.

[24] Nagai H, Takahashi J, Kemmochi K, Matsui J-i. Inventory analysis in production and recycling process of advanced composite materials. Journal of Advanced Science. 2001;13(3):125–128.

[25] Bell J, Pickering S, Yip H, Rudd C. Environmental Aspects of the Use of Carbon Fibre Composites in Vehicles –Recycling and Life Cycle Analysis. End of Life Vehicle Disposal--Technical, Legislation, Economics (ELV 2002). Warwick, UK; 2002.

[26] The Japan Carbon Fiber Manufacturers Association. Carbon fibre reinforced plastic report. 2006.

[27] Hedlund A. Model for End of Life Treatment of Polymer Composite Materials: Royal Institute of Technology; PhD thesis., KTH, 2005.

[28] Carberry W. Airplane Recycling Efforts benefit Boeing operators. Aero. 2008;4(2008):7–13. p. 6–13.

[29] Duflou JR, De Moor J, Verpoest I, Dewulf W. Environmental impact analysis of composite use in car manufacturing. CIRP Annals – Manufacturing Technology. 2009;58(1):9–12.

[30] Song YS, Youn JR, Gutowski TG. Life cycle energy analysis of fiber-reinforced composites. Composites Part A: Applied Science and Manufacturing. 2009;40(8):1257–1265.

[31] Griffing E, Overcash M. Carbon fiber HS from PAN [UIDCarbFibHS]. 1999- present: Chemical Life Cycle Database; 2010.

[32] Zhang X, Yamauchi M, Takahashi J. Life cycle assessment of CFRP in application of automobile. 18 the International Conference on Composite Materials 2011.

[33] Asmatulu E. End-of-life analysis of advanced materials: Wichita State University; PhD thesis., Wichita State University, College of Engineering, Department of Industrial and Manufacturing Engineering, 2013.

[34] Witik RA, Teuscher R, Michaud V, Ludwig C, Manson J-AE. Carbon fibre reinforced composite waste: An environmental assessment of recycling, energy recovery and landfilling. Composites, Part A. 2013;49:89–99.

[35]   Michaud V. Inventory Data of Carbon Fibre, Personal Communication with Prof. Véronique Michaud in Laboratoire de Technologie des Composites et Polymères (LTC), ÉcolePolytechnique FédéraledeLausanne (EPFL) in Switzerland. 2014.

[36]   Schmidt JH, Watson J. Eco Island Ferry: Comparative LCA of island ferry with carbon fibre composite based and steel based structures. In: Consultants L, editor. Aalborg, Denmark 2014.

[37]   Prinçaud M, Aymonier C, Loppinet-Serani A, Perry N, Sonnemann G. Environmental Feasibility of the Recycling of Carbon Fibers from CFRPs by Solvolysis Using Supercritical Water. ACS Sustain Chem Eng. 2014:2(6):1498–1502.

[38]   Overcash Ga. Carbon fiber HS from PAN [UIDCarbFibHS]. Chemical Life Cycle Database 2010.

[39]   Oak Ridge National Laboratory. ORNL seeking U.S. manufacturers to license low-cost carbon fiber process. 2016.

[40]   DOE Office of Energy Efficiency and Renewable Energy. "Clean Energy Manufacturing Innovation Institute for Composite Materials and Structures," Funding Opportunity Announcement (FOA) Number DE-FOA-0000977, issued 2/265/2014. 2014.

[41]   DOE. Chapter 6: Innovating Clean Energy Technologies in Advanced Manufacturing Technology Assessments. Quadrennial Technology Review 2015. 2015 p. 12.

[42]   Meng F, McKechnie J, Turner TA, Pickering SJ. Energy and environmental assessment and reuse of fluidised bed recycled carbon fibres. Compos Part A-Appl S. 2017;100:206–214.

[43]   Witik RA, Teuscher R, Michaud V, Ludwig C, Manson J-AE. Carbon fibre reinforced composite waste: An environmental assessment of recycling, energy recovery and landfilling. Composites Part a-Applied Science and Manufacturing. 2013;49:89–99.

[44]   Patel M. Cumulative energy demand (CED) and cumulative CO2 emissions for products of the organic chemical industry. Energy. 2003;28(7):721–740.

[45]   Gabi. Gabi Extension Database VII Plastics. 2014, https://sphera.com/.

[46]   Wernet G, Bauer C, Steubing B, Reinhard J, Moreno-Ruiz E, Weidema B. The ecoinvent database version 3 (part I): overview and methodology. The International Journal of Life Cycle Assessment. 2016;21(9):1218–1230.

[47]   Duflou JR, Deng Y, Van Acker K, Dewulf W. Do fiber-reinforced polymer composites provide environmentally benign alternatives? A life-cycle-assessment-based study. MRS Bull. 2012;37 (04):374–382.

[48]   Scelsi L, Bonner M, Hodzic A, Soutis C, Wilson C, Scaife R, et al. Potential emissions savings of lightweight composite aircraft components evaluated through life cycle assessment. Express Polym Lett. 2011;5(3):209–217.

[49]   Li X, Bai R, McKechnie J. Environmental and financial performance of mechanical recycling of carbon fibre reinforced polymers and comparison with conventional disposal routes. Journal of Cleaner Production. 2016;127:451–460.

[50]   Shuaib NA, Mativenga PT. Energy demand in mechanical recycling of glass fibre reinforced thermoset plastic composites. J Clean Prod. 2016;120:198–206.

[51]   Shibata K, Nakagawa M. Hitachi Chemical Technical Report: CFRP Recycling Technology Using Depolymerization under Ordinary Pressure. 2014.

[52]   Keith MJ, Oliveux G, Leeke GA. Optimisation of solvolysis for recycling carbon fibre reinforced composites. European Conference on Composite Materials 17. Munich, Germany 2016.

[53]   Meng F. Environmental and cost analysis of carbon fibre composites recycling: University of Nottingham; 2017.

[54]   UK Department for Business Energy & Industrial Strategy (BEIS). Greenhouse gas reporting: conversion factors 2020. 2020.

[55]   Kara M. Environmental and economic advantages associated with the use of RDF in cement kilns. Resour, Conserv Recycl. 2012;68:21–28.

[56] Palmer J, Savage L, Ghita OR, Evans KE. Sheet moulding compound (SMC) from carbon fibre recyclate. Composites, Part A. 2010;41:1232–1237.

[57] Palmer J. Mechanical recycling of automotive composites for use as reinforcement in thermoset composites [PhD thesis]: University of Exeter; 2009.

[58] Liu P, Meng F, Barlow CY. Wind turbine blade end-of-life options: An eco-audit comparison. Journal of Cleaner Production. 2019;212:1268–1281.

[59] Cunliffe AM, Jones N, Williams PT. Recycling of fibre-reinforced polymeric waste by pyrolysis: thermo-gravimetric and bench-scale investigations. Journal of Analytical and Applied Pyrolysis. 2003;70(2):315–338.

[60] Job S. Composite recycling-summary of recent research and development-Materials KTN Report. 2010.

[61] ELG Carbon Fibre Ltd. LCA benefits of rCF. 2017.

[62] Pender K, Yang L. Investigation of the potential for catalysed thermal recycling in glass fibre reinforced polymer composites by using metal oxides. Composites, Part A. 2017;100:285–293.

[63] Pickering SJ, Turner TA, Meng F, Morris CN, Heil JP, Wong KH, et al. Developments in the fluidised bed process for fibre recovery from thermoset composites. CAMX 2015 – Composites and Advanced Materials Expo 2015. p. 2384–2394.

[64] Pickering SJ, Kelly RM, Kennerley JR, Rudd CD, Fenwick NJ. A fluidised-bed process for the recovery of glass fibres from scrap thermoset composites. Composites Science and Technology. 2000;60 (4):509–523.

[65] Pender K, Yang L. Regenerating performance of glass fibre recycled from wind turbine blade. Composites, Part B. 2020;198:108230.

[66] Liu Y, Meng L, Huang Y, Du J. Recycling of carbon/epoxy composites. Journal of Applied Polymer Science. 2004;94(5):1912–1916.

[67] Oliveux G, Dandy LO, Leeke GA. Current status of recycling of fibre reinforced polymers: Review of technologies, reuse and resulting properties. Progress in Materials Science. 2015;72:61–99.

[68] Wheatley A, Warren D, Das S. Low-Cost Carbon Fibre: Applications, Performance and Cost Models. In: Advanced Composite Materials for Automotive Applications, Elmarakbi, A. Ed. 2013:405–434 https://doi.org/10.1002/9781118535288.ch17.

[69] Eberle raf, H. Modelling the Use Phase of Passenger Cars in LCI. SAE Total Life-cycle Conference. Graz Austria: SAE Technical Paper 982179 1998.

[70] Ridge L. EUCAR-automotive LCA guidelines-phase 2. SAE Technical Paper; 1998.

[71] Helms H, Lambrecht U. The potential contribution of light-weighting to reduce transport energy consumption. Int J Life Cycle Ass. 2007;12(1):58–64.

[72] Koffler C, Rohde-Brandenburger K. On the calculation of fuel savings through lightweight design in automotive life cycle assessments. Int J Life Cycle Ass. 2010;15(1):128–135.

[73] Ashby MF. Materials Selection in Mechanical Design(3rd edition). Butterworth-Heinemann, Oxford, UK 2005.

[74] Li F, Patton R, Moghal K. The relationship between weight reduction and force distribution for thin wall structures. Thin-walled structures. 2005;43(4):591–616.

[75] Patton R, Li F, Edwards M. Causes of weight reduction effects of material substitution on constant stiffness components. Thin-Walled Structures. 2004;42(4):613–637.

[76] Wong KH, Pickering SJ, Turner TA, Warrior NA. Compression moulding of a recycled carbon fibre reinforced epoxy composite. SAMPE 2009 Conference. Baltimore, Maryland 2009.

[77] Plasti Comp Inc. 30% Long Carbon Fiber Reinforced PP – Complēt LCF30-PP. 2016, https://www.good fellow.com/uk/material/composites.

[78] Good Fellow. Technical Information – Carbon/Epoxy Composite. 2016.

[79]   Berthelot J-M. Composite Materials: Mechanical Behavior and Structural Analysis. New York: Springer Science & Business Media; 2012.

[80]   Daniel IM, Ishai O, Daniel IM, Daniel I. Engineering Mechanics of Composite Materials: Oxford University Press New York; 1994.

[81]   MatWeb. Technical Data Sheet-AISI 1017 Steel, cold drawn. 2016.

[82]   ASM Aerospace Specification Metals Inc. Aluminum 6061-T6. 2015.

[83]   Nam EK, Giannelli R. Fuel consumption modeling of conventional and advanced technology vehicles in the Physical Emission Rate Estimator (PERE). US Environmental Protection Agency. 2005.

[84]   Kim HC, Wallington TJ, Sullivan JL, Keoleian GA. Life Cycle Assessment of Vehicle Lightweighting: Novel Mathematical Methods to Estimate Use-Phase Fuel Consumption. Environ Sci Technol. 2015;49(16):10209–10216.

Index

https://doi.org/10.1515/9783110754438-005

www.ingramcontent.com/pod-product-compliance
Lightning Source LLC
Chambersburg PA
CBHW061418210326
41598CB00035B/6252